GLOBAL TOURISM DEVELOPMENT

Ann Kenward and Jan Whittington

Hodder & Stoughton

A MEMBER OF THE HODDER HEADLINE GROUP

Acknowledgements

For Roger, Neil and Huw

The authors and the publishers would like to thank the following for permission to reprint copyright photographs in this book: NZ 1.3, NZ 1.4, NZ 1.5, NZ 1.6, NZ 2.1, NZ 4.1, NZ 4.2, 4.5, Mary Hurren; 2.11, 2.13, 2.15, 4.16, S2.2, Life File; 2.22, 2.25, Corbis; 3.4, Helen Wilson; 4.5, Malcolm Robinson; 2.14, Tim Kenward; 2.2, 2.3, Julie Watson.

All other photographs were provided by the authors.

The publishers would like to thank the following for permission to reproduce copyright material in this book: New Zealand Tourist Office, BTA, Office of Statistics, WTO, Wallace Arnold, Mills and Morrison, Spanish National Tourist Board, The Gambia Tourist Board, J. Christopher Holloway, California State Tourism Office, SAGA, Penguin Books, Disney World Resort (Florida). Ordnance Survey, The Times, the Yorkshire Evening Post.

Every effort has been made to contact the holders of copyright material used in this book, but if any have been overlooked the publishers will be pleased to make the necessary alterations at the first opportunity.

Help and information provided from: Elizabeth Bishop, Janice Brown, Betty Magnusson, Dr. Stephen Page, Meryl Palit, Alayne Reeves and Sue Wylan.

Orders: please contact Bookpoint Ltd, 39 Milton Park, Abingdon, Oxon OX14 4TD. Telephone: (44) 01235 400414, Fax: (44) 01235 400454. Lines are open from 9.00–6.00, Monday to Saturday, with a 24 hour message answering service. Email address: orders@bookpoint.co.uk

British Library Cataloguing in Publication Data
A catalogue record for this title is available from The British Library

ISBN 0 340 72119 7

First published 1999
Impression number   10 9 8 7 6 5 4 3 2 1
Year   2004 2003 2002 2001 2000 1999

Copyright © 1999 Kenward, Whittington

All rights reserved. No part of this publication may be reproduced or transmitted in any form or by any means, electronic or mechanical, including photocopy, recording, or any information storage and retrieval system, without permission in writing from the publisher or under license from the Copyright Licensing Agency Limited. Further details of such licenses (for reprographic reproduction) may be obtained from the Copyright Licensing Agency Limited, of 90 Tottenham Court Road, London W1P 9HE

Cover photo from Colorific
Typeset by Fakenham Photosetting Limited, Fakenham Norfolk NR21 8NL
Printed in Great Britain for Hodder & Stoughton Educational, a division of Hodder Headline Plc, 338 Euston Road, London NW1 3BH by Redwood Books, Trowbridge, Wiltshire

Contents

Introduction 2

Section 1: Tourism - A growth Industry 4
Classifications 4
The Evolution of Tourism 9
The Distribution of Tourism 12

Section 2: Tourism and Society 26
Social Issues Related to Tourism 26
Changes in Tourist Travel Patterns 29
Accommodation 32
Amenities 40

Section 3: Tourism and the Economy 53
Tourism as an Industry 53
Suppliers to the Industry 55
Changes with Time 58
The Impact on the Economy and Development 59

Section 4: Tourism and the Environment 71
Landscapes that are exploited 71
The Impact upon Ecosystems and Environments 72
Hot Spots and Honey Pots 77
The Balance between Conservation and Commercialism 80

Section 5: Tourism and Management 93
Different Attitudes and Values towards Tourism 93
Government Intervention 97
Managing the Environment for Tourism 101
Ecotourism and Sustainable Tourist Development 106

Recurring Case Studies

Brighton
Growth 19
Changing Tourist Amenities 44
Business Tourism 65
Pier Conservation 84
Local Government Plans 112

Spain
Growth 20
Growth of Amenities and Accommodation 47
Negative Impacts of Tourism 66
The Impact of Tourism on Built Environments 85
Reinventing Tourism 114

The Gambia
Background 22
Provision of Tourist Accommodation 48
Positive Aspects of Tourism 67

Conservation and Tourism	87
Planning for the Future	117

New Zealand

Background	23
Social Issues Relating to Tourism	51
Supply and Demand	68
Ecotourism	89
Regional Development of Tourism	118

The Disney Concept

History	25
The Impact on the Surroundings	52
Business Ventures	70
Recreating the Natural Environment	91
Future Developments	120

INTRODUCTION

Tourism has become a multi-billion dollar, worldwide industry, contributing about an eighth of the global economy. It affects societies, economies and environments, connecting them in a complex mesh of interrelationships. Attempts to manage its effects are made to a degree in every corner of the globe.

For clarity, this book attempts to separate the strands, examining the effect of tourism on society (Chapter 2), on the economy (Chapter 3) and on the environment (Chapter 4), whilst the final chapter looks at management strategies. Suitable case studies are used throughout to illustrate the points being made.

However, the reader needs to be aware that the reality is not one of neat, separate parcels. Tourism affects the society, economy and environment of a location simultaneously and as any element

CHAPTER	1 TOURISM – A GROWTH INDUSTRY					2 TOURISM AND SOCIETY			
SECTIONS WITHIN EACH CHAPTER	WHAT ARE LEISURE, RECREATION & TOURISM?	CLASSIFICATION OF TOURISM	THE EVOLUTION & EARLY GROWTH OF TOURISM	FACTORS CREATING THE GROWTH OF TOURISM	DISTRIBUTION OF TOURISM	CONFLICTS & BENEFITS WITHIN SOCIETIES & COMMUNITIES FROM TOURISM	CHANGES IN TOURISTS & COMMUNICATION NETWORKS	ACCOMMODATION – PROVISION & DEMAND AND THEIR IMPACT ON COMMUNITIES	AMENITIES – PROVISION & DEMAND AND THEIR IMPACT ON COMMUNITIES
ASPECTS HIGHLIGHTED BY RECURRING CASE STUDIES									
BRIGHTON			★						★
SPAIN				★		★			
THE GAMBIA						★			
NEW ZEALAND				★		★			
THE DISNEY CONCEPT									★
A LEVEL SYLLABI CONTENT COVERED IN THIS BOOK									
LONDON SYLLABUS B			★	★	★	★	★	★	★
CAMBRIDGE MODULAR MODULE 4545		★	★	★	★	★	★	★	★
OXFORD & CAMBRIDGE			★	★	★	★	★	★	★
CAMBRIDGE LINEAR			★		★	★		★	★
AEB	★		★	★	★	★	★	★	★
WELSH	★	★	★	★		★	★	★	★

FIGURE 1

4 GLOBAL TOURISM

changes, then, like a ripple, the consequences can be far-reaching and unforseen.

These aspects of tourism form the core of the various leisure and tourism units in the A Level examination syllabi. Figure 1 shows how the various components of the syllabi are contained within this book.

There are five major case studies recurring throughout the book to illustrate the interrelationships. These are:

- Brighton (an historic English seaside resort)
- The Spanish 'Costas' (the result of mass tourism in an MEDC)
- The Gambia (an LEDC)
- New Zealand (relying mainly upon the attraction of its physical environment) centre
- The Disney Concept (a purpose-built environment and entertainment)

The themes presented in each chapter are examined in turn for each of these five locations. It is hoped that in this way, the student can see how the complex relationships build up to form an overall picture, and at the same time gain a broad knowledge of five disparate areas, which can be applied to any examination question as case study material. Figure 1 also shows what aspect of each chapter is highlighted by these recurring case studies.

	3 TOURISM AND THE ECONOMY					4 TOURISM AND THE ENVIRONMENT				5 TOURISM AND MANAGEMENT				
	TOURISM AS AN INDUSTRY SUBJECT TO THE LAWS OF SUPPLY & DEMAND	TOURISM SUPPLIERS	TOURISM AND LEVELS OF DEVELOPMENT AT DIFFERENT SCALES	TOURISM AND ITS IMPACT ON ECONOMIES AT DIFFERENT SCALES	MODELS OF GROWTH	THE EFFECT OF TOURISM ON ECOSYSTEMS	ENVIRONMENTS EXPLOITED BY TOURISTS & CONFLICTS ARISING	HOTSPOTS & HONEY POTS – UNEVEN DEVELOPMENT	CONSERVATION VS. COMMERCIALISM	THE IMPACT OF DIFFERENT PHILOSOPHIES	LEVELS OF MANAGEMENT AND THEIR IMPACT	INEQUALITIES IN DEVELOPMENT NORTH & SOUTH	TOURISM MANAGEMENT AND THE ENVIRONMENT	ECOTOURISM AND SUSTAINABLE TOURIST DEVELOPMENT
			★						★	★				
			★						★	★				
			★				★					★		
			★			★								★
	★								★	★				
	★		★	★	★	★	★	★	★	★	★	★	★	★
			★	★		★	★	★	★	★	★	★	★	★
			★	★			★	★	★	★	★		★	
			★	★		★	★	★					★	
			★	★			★	★	★				★	★
	★	★	★	★		★	★	★	★	★	★	★	★	

1
TOURISM — A GROWTH INDUSTRY

Key Ideas

- Tourism may be classified according to several criteria, many of which are interrelated
- Tourism has evolved and developed over time
- The substantial expansion in the demand for recreation and tourism has resulted from a variety of factors including growth in incomes, leisure time and mobility
- Tourism growth has resulted in significant national and international flows of services, goods and people

What are Leisure, Recreation and Tourism?

These three words – leisure, recreation and tourism – are often used in the same sentence and to many people seem to be interchangeable. Do they really mean the same thing? The Oxford English dictionary suggests subtle differences:

- Leisure is the freedom or opportunity to do something, the state of having time at one's disposal.
- Recreation is a pleasant occupation, pastime or amusement.
- Tourism is travelling for pleasure.

From this we could say that **leisure** is a general term for what we do when we are not working, studying or any other similar enforced activity. We use this free time in a variety of ways which may include recreation or tourism. The term **recreation** can include both sporting activities and more sedentary pastimes such as playing electronic games, although it does not include sitting around relaxing and other similar 'non-activities'.

Academics have produced a huge quantity of words in an attempt to define tourism, often contradicting each other. However, they mainly agree that **tourism** involves travelling away from home for longer than 24 hours. (Note that the purpose of the trip is not included – although many people associate the term tourism with holidays and therefore pleasure and leisure, travelling away from home on business is considered a form of tourism but not a leisure activity).

With some thought, it is possible to produce examples of activities which fit neatly into these definitions. A little further thought produces examples which blur the boundaries between them. For example, playing tennis at a local park is obviously a recreational pastime (sport) whilst staying at home and watching an international tennis tournament on television is a general leisure activity. A game of tennis whilst on holiday is both recreation and tourism. Figure 1.1 illustrates these overlapping relationships.

Leisure in all its forms is a major world industry. Figure 1.2 (which defines tourism as any trip of 40 or more km away from home) shows the economic force of the tourist industry. It is responsible for approximately one eighth of world earnings.

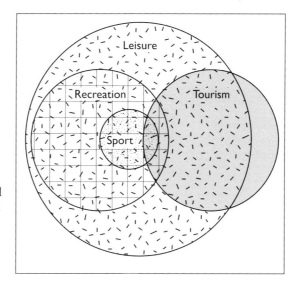

FIGURE 1.1 The relationship between Leisure, Recreation and Tourism

FIGURE 1.2 Tourism as a billion dollar industry

	1990 (US$ billion)	2000	% Real average Annual increase
Total World Receipts from Domestic and International Tourism	2660	4330	5.0
World Receipts from International Tourism (incl. Transport)	319	519	5.0
World Receipts from International Tourism (excl. Transport)	255	415	5.0
World Receipts from Domestic Tourism (incl. Transport)	2341	3811	5.0
Total World Airline Passenger Revenues	211	343	5.0
Worldwide Accommodations Revenue	222	301	3.1
	(US$ trillion)		
World Gross Product	22.4	33.1	4.0
Ratio – Total World Tourism Receipts to World Gross Products	12%	13%	–

NB. Estimates are based on trips 40 km or more from home.
Figures exclude inflation and exchange rate fluctuations.

Source: Travel Industry World Yearbook

The Classification of Tourism

Tourism can be classified by resident origin:
a) **Domestic Tourism** – when a person visits another place in his or her own country.
b) **Inbound Tourism** – visits to a country by non-residents of that country.
c) **Outbound Tourism** – residents of one country visiting destinations in another country.

These form sub-groups of a broader classification which is shown in Figure 1.3:
1 **Internal Tourism**, which consists of domestic and inbound tourists
2 **National Tourism**, which consists of domestic and outbound tourists
3 **International Tourism**, which consists of inbound and outbound tourists.

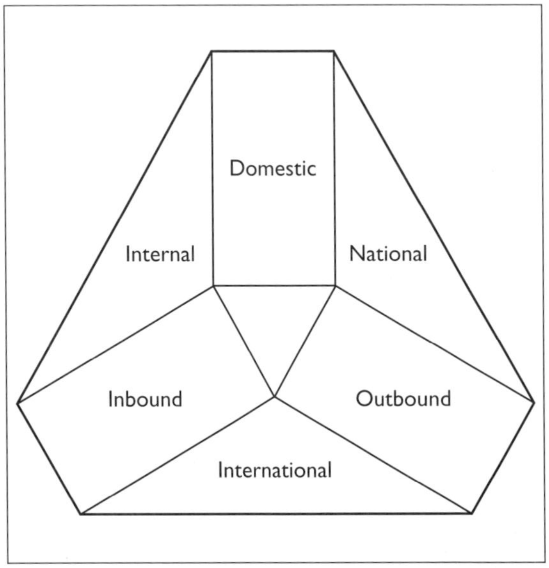

Other methods of classification of tourism consider motivation, destination, modes of travel and organisation.

People travel for a wide variety of reasons to a large number of destinations. They use several methods of transport, are away for different lengths of time at various times of the year. Their journeys and the facilities which they use at their destinations are organised in a variety of ways. These factors combine to form complex interrelationships but to aid understanding each one will be considered separately.

Why do tourists travel?

The travel industry identifies the following motivations:

Relaxation and physical recreation
Some people travel to coastal, countryside, mountainous or wilderness areas to commune with nature or for sun, sea and sand.

Touring, sightseeing and culture
Many people prefer to travel from one place to another while on holiday; in particular to see the built environment and to experience other people's culture and lifestyles.

Visiting friends and relatives (known in the tourist trade as 'VFR Tourism')
A large percentage of people spend holiday times with friends and relations. Their destination is often very different from a 'normal' tourist location.

Business trips and conferences
These are city-oriented and characterised by their short length and all-year-round occurrence. Conferences have recently become a major economic factor in the tourist industry.

Specialist trips
These are taken for health, study, sporting or religious reasons.

FIGURE 1.3 Classification of Tourism (after Steven Smith)

These motivations for travel are illustrated by Figure 1.4 which gives details of UK visitors to New Zealand over a five year period.

Where do tourists go?

Tourist destinations depend upon

- their perceived **attractions** for the tourist;
- their **amenities** or facilities;
- their **accessibility** in terms of travel time and distance involved.

Attractions
These may be natural or man-made but the more successful tourist destinations tend to offer a combination of both. In this context, weather may be seen as a natural attraction.

Amenities
Tourists need certain basic amenities such as accommodation and food. They will also want local transport and entertainments at the site. The presence of these facilities may enhance the original feature or detract from it.

Accessibility
Areas which are readily accessible have regular and convenient forms of transport at a reasonable price.

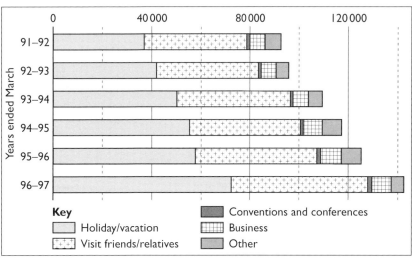

FIGURE 1.4 UK visitors to New Zealand 1991–96 (New Zealand Tourist Board)

How do tourists travel?

By definition travel is an integral part of the tourist industry. Tourists use all modes of transport i.e. road, rail, water and air to reach their destinations and to explore the surrounding area on arrival. The growth of tourism has always been closely linked with developments in transport. This relationship is explored more fully later in this chapter in the section on factors influencing the growth of tourism.

STUDENT ACTIVITY 1.1

1 Using Figure 1.1 give an example of each activity – leisure, recreation, sport and tourism (including overlaps)
(a) which you have taken part in and
(b) which are available in your local area.

2 Using sources of information about New Zealand e.g. tourist brochures, CD ROMs, Encyclopaedia, suggest what New Zealand's main attractions will be for those travellers visiting for:
a) relaxation and physical recreation and
b) tourism, sightseeing and culture.

3 Identify, in your home area within a distance of 100 km, places suitable for day trips, weekend visits and a week's holiday.
a) Collect information, for example from brochures and advertising material, about amenities provided in each of your locations and write a summary paragraph on each.
b) Identify any differences in provision of amenities that each length of visit requires.

4 With reference to the information shown on Figures 1.5a and 1.5b write an account to explain how travel to holiday destinations has changed during the last 20 years. You could take each line of the table in turn and think about reasons why the proportion of holiday makers using that mode of transport or that airport has changed.

FIGURE 1.5a Transport used by British holidaymakers to travel abroad

	Plane %	Boat %	Hovercraft %	Channel Tunnel
1970	61	36	3	–
1972	69	27	4	–
1974	72	26	3	–
1976	73	27	3	–
1978	69	27	3	–
1980	68	28	4	–
1982	66	30	3	–
1983	67	31	2	–
1984	70	27	2	–
1985	70	28	2	–
1986	71	27	1	–
1987	76	23	1	–
1988	77	21	2	–
1989	76	21	2	–
1990	73	26	1	–
1991	74	27	2	–
1992	76	24	<0.5	–
1993	76	22	1	–
1994	80	19	<0.5	–
1995	78	18	<0.5	2
1996	78	15	<0.5	5

NB Figures may not add up to 100% due to rounding
(Source: BTA National Travel Survey)

FIGURE 1.5b Traffic at selected airports in Britain

	Million passengers			
	1970	1980	1990	1996
London Heathrow	15.6	27.5	43.0	55.8
London Gatwick	3.7	9.7	21.2	24.1
Manchester	1.9	4.3	10.1	14.8

(Source: EU transport in Britain (statistical pocket book))

8 GLOBAL TOURISM

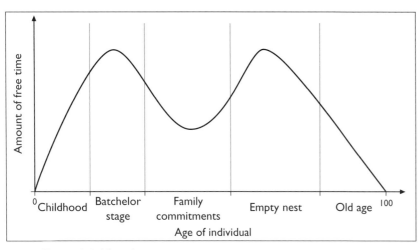

FIGURE 1.6 (above) The impact of age on leisure time

How long are tourists away?

The length of time a tourist spends away from home may depend upon:

- age – a simple relationship between a person's age and the time available for holidays is shown in Figure 1.6. Most holidays are taken by young adults and middle-aged people who have few financial burdens and family commitments.
- employment – the amount of paid holiday entitlement has increased steadily throughout the twentieth century. Its impact is considered more fully in the section of this chapter concerning factors influencing the growth of tourism. The length of a business trip is largely dependent upon the nature of the business involved.

	Weekly amount spent (£) according to employment (1995–96)						
	Professional	Employers/ managers	Intermediate non-manual	Junior non-manual	Skilled manual	Semi-skilled manual	Unskilled manual
Goods	24.65	21.17	18.58	11.74	16.52	12.99	10.21
Services	55.87	58.89	38.21	31.21	29.58	23.79	20.34

(Source: Office of Statistics)

FIGURE 1.7a (above) Average weekly expenditure on leisure goods and services, according to employment

FIGURE 1.7b Average weekly expenditure on leisure goods and services, by age leaving school

Age leaving school	Expenditure (£) leisure goods	Expenditure (£) leisure services (1995–96)
<15	6.86	18.30
15	12.86	29.25
16	14.52	30.28
17–19	16.87	43.36
19–22	20.29	49.73
22+	23.69	54.08

(Source: Office of Statistics)

- disposable income – this is the amount of money a person has left after meeting his or her financial commitments. In order to travel, a tourist needs money. The amount of disposable income people have available for spending on a holiday is closely related to the nature of their employment (Figure 1.7a), their level of education (as expressed by age upon completing education (Figure 1.7b)) and their total income (Figure 1.7c).

FIGURE 1.7c Average weekly household expenditure

	Weekly household expenditure (£) in 1995–96		
	lowest 10%	average	highest 10%
Housing	12.35	48.25	102.48
Fuel	9.01	12.92	17.39
Food	22.04	52.88	93.97
Alcohol	2.40	11.41	24.74
Tobacco	3.80	5.81	5.19
Clothing/Footwear	3.86	17.15	40.32
Household Goods	7.72	23.45	48.55
Household Services	5.26	15.13	37.15
Personal Goods/Services	3.42	11.55	24.33
Motoring	5.69	36.99	83.16
Fares/Travel	2.15	6.17	15.79
Leisure Goods	3.09	13.73	32.58
Leisure Services	6.59	32.05	91.82
Miscellaneous	0.32	2.37	5.94
Total	87.80	289.86	623.42

(Source: Office of Statistics)

TOURISM – A GROWTH INDUSTRY

Country	GNP $ per capita	Population (millions)	Numbers outbound (thousands)	% Population who travel abroad
Antigua	6980	0.07	395	564
Brazil	3640	155	2596	1.6
China	960	1,221.46	4520	0.3
France	24 990	58.14	18 686	32
Belgium	24 170	10.14	6453	63
The Gambia	320	1.12	n/a	
Ireland	14 710	3.58	2368	66
Netherlands	24 000	15.45	1026	7
New Zealand	14 340	3.54	920	26
Portugal	9470	10.8	227	2
Singapore	26 730	2.99	2867	95
Zimbabwe	540	11.53	256	2.2
Spain	13 580	39.21	7708	20
Great Britain	18 700	58.2	41 873	72

FIGURE 1.8 GNP for a range of countries, and the numbers of their population travelling abroad in 1995

(Source: WTO)

STUDENT ACTIVITY 1.2

1 For Figure 1.7a, describe the average weekly expenditure on leisure goods and suggest a leisure activity and typical holiday for each of the employment groups.
2 For Figure 1.7b, suggest why expenditure increases with age at which people leave school.
3 a) For Figure 1.7c, choose one of the columns of figures and calculate the totals for the following combined categories:
– Housing related expenditure
– Food, clothing, personal goods and miscellaneous expenditure
– Non-essential consumption expenditure
– Travel expenditure
– Leisure expenditure.
Draw a pie chart to show your totals.
b) Compare your pie chart with those drawn for the other expenditure groups (remaining columns) and comment upon the differences.
4 With reference to the data in Figure 1.8, calculate Spearman's rank correlation coefficient to test whether there is a relationship between a country's GNP and the percentage of its population who travel abroad.

When do tourists travel?

There are several factors which influence the time of year in which people take a holiday.

- Weather – traditionally people have taken their main holiday in the summer months of the hemisphere in which they live. A much smaller group have chosen to travel in winter for snow-related holidays.
- Public holidays – Christmas and New Year are peak travel times globally, even in non-Christian countries.
- School and College holidays – many families try to keep disruption of their children's education to a minimum by taking their holiday during the closure of educational establishments.

Costa de la luz HOTELS					ISTEL 76# Fastrak ISS	
Hotel Accom code Board		Hotel Costa Golf 5605 HB		Iberostar Andalus Palace 5606 BB		
No of Nights		7	Extra night	7	Extra night	
Jan 2–Feb 28		419	24	429	27	
Mar 1–Mar 30		425	24	439	27	
Apr 1–Apr 13		729	51	569	27	
Apr 14–May 21		495	30	469	28	
May 22–May 28		575	30	585	28	
May 29–Jun 25		545	30	575	41	
Jun 26–Jul 9		639	51	589	41	
Jul 10–Jul 16		876	65	605	41	
Jul 17–Aug 13		865	65	779	54	
Aug 14–Aug 27		839	65	769	54	
Aug 28–Sep 10		695	51	689	54	
Sep 11–Sep 24		675	51	685	54	
Sep 25–Oct 15		519	30	535	28	
Oct 16–Oct 31		509	30	499	28	
Nov 1–Dec 12 & Jan 1–Mar 25		449	26	465	29	
Dec 13–Dec 31		675	38	705	42	
SUPPLEMENTS (per person per night)		Single use of suite: £11		Half board: £5 Single use of twin: £13 Junior suite: £20		
FLIGHT SUPPLEMENTS See page 196/7 for details		Supplement		Prices are per person sharing a twin room and include UK government air passenger duty of £10.		
Gatwick–Jerez BA Mon, Wed, Fri		Nil				
British Airways Club Class supplement: £225				Transfers are included. Transfer time: approx 3/4 hour. For holiday insurance, booking conditions and our price policy, all of which must be read before booking, please see pages 197–201.		

FIGURE 1.9 Holiday brochure prices

- Factory closures – some places of employment close down at certain times of the year. These times usually coincide with school or public holidays for the convenience of the employees.
- Work rotas – to maintain production or provision of services a work organisation may require its employees to take their holiday entitlement at specific times during the year.

These factors in combination are responsible for creating certain times of year when supply is exceeded by demand and higher prices result. (See Figure 1.9). Some destinations are in demand all-year-round whilst others suffer from **seasonality**.

Who organises the holiday?

People may organise all aspects of their holiday independently, purchase some elements from a tour operator or buy a complete **package**. The introduction and growth of the package holiday has played a major role in the growth of world tourism over the past 30 years. Initially, package holidays provided full board and transport. Today there is a growing trend for self-catering accommodation reflecting increased confidence and desire for independence by tourists. This is shown in Figure 1.10.

FIGURE 1.10 Changes in accommodation used in holidays abroad and in Great Britain since 1971

	% Holidays at home						% Holidays abroad					
	1971	1976	1981	1986	1991	1996	1971	1976	1981	1986	1991	1996
Hotel	17	16	17	20	19	20	66	60	58	55	47	46
Boarding House	18	11	7	7	6	5	4	3	3	3	3	3
VFR	27	25	26	23	19	17	20	24	21	16	20	21
Caravan	19	21	18	21	23	22	2	2	3	3	3	2
Rented Accom.	10	11	14	13	14	13	3	5	9	18	23	22
Holiday Camp	6	6	6	9	8	10	1	1	1	1	1	1
Camping	8	7	7	7	8	5	7	6	6	3	3	2
Youth Hostel	data not available						1	1	1	1	1	2
Boats & cruises	data not available						4	2	2	2	3	4
Other	4	4	5	5	7	8	2	5	4	7	9	10

(Source: British National Travel Survey)

The Evolution and Early Growth of Tourism

'The use of travelling is to regulate imagination by reality and instead of thinking how things may be, to see them as they are'

Samuel Johnson

Before the Industrial Revolution

During the late seventeenth and early eighteenth century it became fashionable for young members of the nobility to complete their education by taking the 'Grand Tour' of Europe. The journey generally began in London, moved on to Paris, and then to Italy with stops in Genoa, Milan, Florence, Rome and Venice. The return trip went through Switzerland, Germany and The Netherlands. Boswell, in his biography of Dr. Johnson, summed up this activity with the words:

'A man who has not been to Italy is always conscious of an inferiority, from his not having seen what it is expected a man should see ... the grand object of travelling is to see the shrines of the Mediterranean.'

Medical opinion at this time claimed that mineral-rich waters and hot springs would be beneficial. Only the rich and powerful could take advantage of this advice so places like Bath and Baden-Baden became great attractions for the upper classes and the wealthier merchants. Accommodation and entertainment facilities quickly developed to cater for visitors and became attractions in their own right at these inland spas. During the second half of the eighteenth century there were at least 100 spas in Great Britain and far more in continental Europe. The health benefits of sea bathing and drinking sea water were also published and this led to the development of resorts such as Brighton (see recurring case studies).

During the eighteenth century growing trade and commerce were producing a new 'middle class' in society – a group of people who had money, leisure and a desire to impress others. They built large country houses designed to receive visitors on a mass scale, in order to establish their owners' position on the social ladder. Amongst the most popular houses for visitors in the eighteenth century were Blenheim Palace, Castle Howard and Chatsworth, all of which figure prominently in modern statistics for visitors to stately homes.

Industrial Revolution to the Second World War

This period saw major changes in the scale and type of tourist development. The middle classes were starting to take annual holidays to escape from their responsibilities and the crowded urban environment, seeking relaxation in areas of natural beauty. The working classes were receiving higher wages and more regular working hours in the new industries than they had ever had as land labourers but were still unable to afford much more than day trips.

Steamers on the major rivers provided reliable and cheap transport and day-trip cruises became a popular activity. This led to the growth of coastal resorts near to large industrial towns. Margate and Southend served London for example, whilst Rhyl and Llandudno in North Wales were popular destinations for Liverpudlians ferried across the Mersey. However, the railways had the biggest impact upon tourism. A rail link between an industrial town and a community in its rural hinterland promoted rapid growth. Whitby turned from small fishing village into a popular Victorian resort for West Yorkshire when a rail link was established from York. Tourism as a small business for the upper reaches of society was transformed into the beginnings of the huge industry it is today.

This rapid expansion created the opportunity for professional holiday consultants to set up business. The first and most famous of these is Thomas Cook. His first organised excursion was a train trip from Leicester to Loughborough on 5 July 1841. The train carried 570 passengers, each paying 1 shilling (5 pence) for the privilege. Ten years later, Cook was arranging the transport and lodgings for 165 000 visitors to the World Exposition at Crystal Palace. By 1866 he had organised his first American Tour and in 1874, Cook introduced 'circular notes' which were accepted by banks, hotels, shops and restaurants. They were effectively the world's first travellers' cheques.

The development of tourism continued to be associated closely with railways until after the Second World War.

Post War

At the end of the Second World War there were large numbers of transport aircraft available in North America and Europe which could be purchased cheaply by companies such as Thomas Cook, who wanted to restart their tourism business. Continued jet engine developments allowed large numbers of people to travel long distances at high speed so that by the 1960s everywhere was easily accessible from everywhere else.

Factors Creating Tourist Growth in The Latter Half of the Twentieth Century

'What we do in our working time determines what we have; what we do in our leisure hours determines what we are'

George Eastham in Readers' Digest

The reasons why there has been such an immense growth in the number of tourists since the Second World War are numerous and interrelated. The ways in which people live their lives have changed dramatically as patterns of work, leisure, income and life style have evolved in response to changes in society. During this period there has also been a revolution in communications, both in the form of transport and in information technology. A major service industry, now employing as many as 10 per cent of the global workforce either directly or indirectly, has grown to meet this phenomenal demand from tourists. However, this industry also markets itself aggressively and in doing so creates further demand for its product.

Changing Living Standards in MEDCs

In the last 40 years there have been major changes in living standards throughout the world. As countries have developed so their inhabitants have received an increased amount of disposable income, that is money not required to purchase the essentials of food and housing. Thus more money can be spent on travel and holidays (see Figure 1.7). Conditions of employment have also improved as many people now have several weeks of **paid holiday** each year. They also have a **shorter working week**. The average number of hours worked in Britain has decreased from 42 hours a week in 1956 to 38 hours in 1995 (see Figure 2.20). It is now accepted that most women work, so that many households are classified as 'dual income'. **Labour saving devices** mean that less time is spent on household chores than formerly. **Commuter travel** has increased to a point where a daily journey to work is now accepted as the norm and holiday travel is seen as an extension of this pattern.

Changing Social Patterns

The **communications revolution** of the last 40 years means that we now live in a global village in terms of the ease with which we can see and hear what is going on anywhere in the world. Ease of travel has enabled people to move away from where they were born, many to other countries as migrant workers. Family patterns are also changing as people have fewer children and expect to lead lives of self-fulfilment rather than focusing solely on work and child-rearing. **Increasing life expectancy**

FIGURE 1.12 The Orient Express – Luxury train travel

enables people in the developed world to enjoy several decades of life without family responsibilities once their children have grown up. **Early retirement** and state and personal **pensions** have provided the time and money for increased travel by members of this group, often called 'empty nesters' in the travel trade! While these changing social patterns have occurred rapidly in the western world, there is also a softening of traditional views in Asia and the Far East, giving rise to increased tourist demand in those areas.

Developments in transport technology

Travel by Rail

It was the invention of the railway that led to the growth of mass holidays in the nineteenth century

FIGURE 1.11 Changing modes of holiday transport in Great Britain

	1971	1976	1981	1986	1991	1996
Car	69	69	68	72	78	75
Bus/Coach	15	13	15	15	12	12
Train	13	11	12	8	7	7
Other	3	4	3	4	4	4

(Source: British National Travel Survey)

and this pattern of travel continued until the end of the Second World War. In the 1950s the number of people travelling by train to their holiday destination fell rapidly. Private car ownership increased dramatically during the 1950s and 60s and it was obviously more convenient, cheaper and easier to travel 'door to door' in one's own time than cope with luggage, crowds and restricted travelling times on the railway. Figure 1.11 shows how some modes of transport have changed in percentage over a 25 year period.

Today those people who travel by train on holiday divide into two main groups. Firstly there are the concession travellers. These groups, especially the elderly and the young, may not own cars and are encouraged to use trains by the offer of special rates. Secondly, there is a large number of rail enthusiasts and nostalgia tourists who enjoy travelling on the great routes of the past, where luxury service is still available, and for whom the journey is the main point of the holiday. Figure 1.12 shows the Orient Express.

Many rail companies are now promoting their advantages over other forms of holiday transport and are winning back tourists from the roads and aeroplanes. A survey of AMTRAK (US Rail) passengers refer to these reasons for choosing rail travel:

- Safety;
- Ability to look out and watch the area through which the train is passing;
- Ability to get up and walk around;
- Arrival at the destination rested and relaxed;
- Personal comfort.

Plans for a fixed-link between Britain and the continent have been drawn since Napoleonic times. It was not until the 1980s that a company, Eurotunnel, was formed and work started (in 1987) to excavate three subterranean tunnels between England and France. These were to contain a railway line in each direction with a service tunnel between. They were completed and brought into operation in 1994. As a result the ferry companies operating channel routes have reduced in number and have improved their road access, terminal facilities and capacity. The *Pride of Dover* – the P&O Stenaline flagship between Dover and Calais – can take 650 cars and 2300 passengers. However, transit time by ferry is still 75 minutes as opposed to a 35 minute journey via the tunnel. Fast-speed hovercraft

TOURISM – A GROWTH INDUSTRY

	Kent–Pas de Calais		Eastern England–Holland/Belgium		Portsmouth/Plymouth–Brittany/Spain		Euro Tunnel		Total		Day trippers
	000s	%	000s	%	000s	%	000s	%	000s	%	%
1992	20 966	70.8	3466	11.7	5197	17.5	–	–	29 629	100	18.8
1993	21 864	70.3	3437	11.1	5782	18.6	–	–	31 083	100	23.6
1994	24 780	73.4	3480	10.3	5087	15.1	411	1.2	33 758	100	25.9
1995	22 522	58.1	3360	8.7	5071	13.1	7797	20.1	38 750	100	29.9
1996	22 909	52.0	2998	6.8	4673	10.6	13 478	30.6	44 058	100	36.2

(Source: Mintel)

FIGURE 1.13 Number of passengers crossing the Channel 1992–1996

By Air (000s)	By Sea/Channel Tunnel (000s)				
	with car	with coach	other	Irish Sea	all
28 624	6374	2657	2633	1586	13 249

(Source: Transport Statistics: Great Britain, 1997)

FIGURE 1.14 Forms of holiday transport for outbound UK passengers, 1995

and catamarans do the crossing in 40 minutes but can only take a limited number of passengers.

The construction of the tunnel between Folkestone and Calais has already altered the pattern of travel between England and France. Figure 1.13 shows the numbers and route of the UK residents who crossed the channel by sea or by rail. It also gives the percentage of day-trippers in those totals. Abolition of duty free concessions in the European Union in the future will almost certainly bring about further changes in travel habits, as will the ultimate impact of monetary union between European states.

Travel by Ship

Ship travel reached its peak in 1957 when 1 036 000 passengers took ocean liners as a means of getting from one continent to another. Then advances in aviation technology, in particular the development of long-haul aircraft, led to a rapid decrease in demand for a slow crossing from America to Europe, or wherever, as passengers took to the skies. The delights of shipboard life were recaptured as special cruise ships were commissioned and this became very popular for middle-aged and elderly people who wanted to travel and to retain their home comforts! In America, where these ships are owned by the major tour operators, the average age of the clientèle is much younger and the range of activities is adjusted accordingly. The major cruise destinations have always been the Mediterranean Sea and the West Indies for ships operating from Europe and North America respectively. Recently a new market has opened up in South East Asia and the Pacific, catering for Japanese and Australian tourists.

Changes in ship design led to the development of 'ro-ro' (roll on, roll off) ferries in the 1970s. This enabled private car owners to travel with their cars and so increased the dominance of the car as a means of holiday transport particularly for island countries such as Great Britain. Figure 1.14 shows the percentage breakdown for each means of holiday transport in Great Britain in 1995.

Travel by Car

Nearly 70 per cent of British holidaymakers use their cars as a means of transport for their holiday, while in the USA the figure is 75 per cent and in Canada it is 85 per cent, although these countries are much larger and the distances travelled are much further.

FIGURE 1.15 Wallace Arnold Coach Tours brochure

FIGURE 1.16 Increase in Air Travel

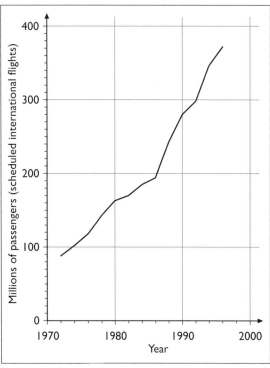

(Source: WTO)

FIGURE 1.17 Range of holiday options

Developed Resorts	with hotels, facilities, sun and entertainment
Beach holidays	families with small children. Sun
Culture and comfort	historic sites; museums and art galleries and good hotels
Sports and entertainment	sports related activity holidays with night life – clubs, casinos, gambling
Culture and Nature	historical sites, museums and art galleries; Wildlife National parks and forests, budget accommodation, camping grounds and trailer parks
Big City	hotels, night life, entertainment. Amusement parks, shopping, restaurants, theatre, sightseeing
Outdoor sports	Strenuous outdoor activity. Hiking, water sports, climbing, skiing

From Mill & Morrison: Tourism product segments

A review of the reasons why people chose to go away by car showed the main factors were:

- Control of the route and of stops en route;
- Control of departure times;
- Ability to carry baggage and equipment easily;
- Low out-of-pocket expenses when travelling with three or more people;
- Freedom to use the car once the destination is reached;
- Enjoyment of driving as a recreational pastime in itself.

This popularity has resulted in a wide range of holidays based on car travel. Different forms of vehicles have also been invented to meet the demand for a complete holiday on wheels. Caravans, motor-homes and trailers are used by a wide range of people for a variety of reasons; the main one is that they wish to tailor their holiday to their car.

Travel by Coach

Another traditional mode of transport is by coach. Travel by 'charabanc' was first introduced for day trips in the 1920s and has always retained a share of the holiday market. Bus companies have lost many passengers as the popularity of the car increased, particularly in respect of short-distance journeys. Long distance travel and the combined coach/accommodation holiday have retained their share of their market especially for the elderly. Figure 1.15 shows a typical advertisement for this type of holiday.

Travel by Air

The last 30 years have seen an enormous expansion in holiday air travel throughout the world. This has been made possible by developments in aviation technology as planes have become larger and faster. Other factors include:

- Provision of aviation infrastructure – runways and airports have been built by governments wishing to attract more tourists;
- Improved air-traffic control procedures allowing almost continuous take-off and landing at busy airports;
- Gradual price de-regulation leading to cheaper fares;
- The growth of package tours in which travel and accommodation were marketed together by tour operators.

Figure 1.16 shows the increase in air travel over the last thirty years.

In general, the mode of transport used by holiday-makers is determined by their choice of destination. Other factors include comfort, cost, speed and convenience as well as the availability and frequency of alternative methods of travel.

TOURISM – A GROWTH INDUSTRY

The emergence of tourist agencies

This explosion in demand for tourist services, whether for transport, accommodation or entertainment has led to the growth of a global industry. Holidays are now very different from the annual week at the seaside taken by the middle classes in Britain in the early 1900s. The range of options offered to the individual as he or she decides what sort of holidays to take is summarised in Figure 1.17.

This shows that for each sort of holiday and for each type of person there is a business organisation waiting to provide what is required. In the last 40 years tour operators and transport firms have grown, merged, collapsed and re-emerged as they have fought each other for a share of a very lucrative market. Governments and international agencies such as the International Monetary Fund are also very aware of the potential for economic growth through tourism and encourage its expansion wherever possible.

> **STUDENT ACTIVITY 1.3**
>
> 1 Summarise the changes in travel patterns prior to 1950. What are the key elements responsible for these changes?
> 2 List the factors influencing tourist growth in the last 50 years. Discuss which ones you feel have made the most important contribution to the growth of tourism.
> 3 Compare the data given in Figure 1.13 for travellers to Europe by ferry and Eurostar (Euro Tunnel). Suggest reasons for any differences you note.
> 4 Consider each form of transport used by holiday makers under the following headings: 'Comfort', 'Cost', 'Speed', 'Convenience', 'Competition'.
> 5 From Figure 1.15, select the elements of the advertisement which will appeal to elderly customers.
> 6 Using Figure 1.17, suggest a range of locations in the British Isles for each of the holiday products listed.

The Distribution of Tourism

The expansion of tourism has led, by definition to a similar expansion in travel, both domestic and international. This in turn has had far-reaching effects on the patterns and flows of goods, services and people throughout the world. An analysis of these flows and patterns can help to visualise the impact of tourism, and help to explain the unevenness of its distribution.

Patterns of international tourism

Tourism is not dispersed evenly throughout the world. Figure 1.18 shows the world's top 15 destinations. Each individual country's rank within the top 15 has varied slightly over the past few years but there have been few additions or subtractions to these basic 15 countries. This spatial variation in the distribution of international tourist activities is influenced by a number of factors.

Rank				Country	International tourist arrivals (thousands)			% Change	Market share % of world total	
1990	1995	1996	1997		1990	1996	1997	97/96	1990	1997
1	1	1	1	France	52 497	62 406	66 800	7.04	11.47	10.90
2	2	2	2	United States	39 363	46 489	49 038	5.48	8.60	8.00
3	3	3	3	Spain	34 085	40 541	43 403	7.06	7.45	7.08
4	4	4	4	Italy	26 679	32 853	34 087	3.76	5.83	5.56
7	5	5	5	United Kingdom	18 013	25 293	25 960	2.64	3.94	4.24
12	8	6	6	China	10 484	22 765	23 770	4.41	2.29	3.88
27	9	9	7	Poland	3 400	19 410	19 514	0.54	0.74	3.18
8	7	7	8	Mexico	17 176	21 405	18 667	−12.79	3.75	3.05
10	11	10	9	Canada	15 209	17 329	17 610	1.62	3.32	2.87
16	12	12	10	Czech Republic	7 278	17 000	17 400	2.35	1.59	2.84
5	6	8	11	Hungary	20 510	20 674	17 248	−16.57	4.48	2.81
6	10	11	12	Austria	19 011	17 090	16 642	−2.62	4.15	2.72
9	13	13	13	Germany	17 045	15 205	15 828	4.10	3.72	2.58
–	18	14	14	Russian Fed.		14 587	15 350	5.23	–	2.50
11	14	16	15	Switzerland	13 200	10 600	11 077	4.50	2.88	1.81

(Source: WTO)

FIGURE 1.18 World's top tourism destinations

GLOBAL TOURISM

The distribution of tourism resources

Tourist resources can be divided into several categories:

- Natural resources – Switzerland was the first destination to develop a tourist industry because of the quality of its landscape. The mountains, lakes, rivers and forests of the Alps proved to be very attractive to wealthy travellers from Northern Europe at the end of the last century.
- Man-made resources – the tourist industry needs large sums of capital expenditure to provide the essential infrastructure for any growth above a basic level. Airports, runways, motorways must be built to bring in the clients who then require accommodation at a variety of levels and entertainments to suit different tastes. Most of the Andean countries in Latin America are just becoming aware of their tourist potential but have yet to build the infrastructure needed. On the other hand, France – the top ranked destination – has an enormous variety of tourist facilities throughout the country.
- Cultural resources – Italy was the first country to attract tourists who came to see the buildings, paintings and townscapes of the Renaissance period. In the twentieth century, our appreciation of our past has extended to include such diverse locations as the remnants of the Aztec civilisation in Mexico and Maori culture in New Zealand.

The wide variety of activities in which travellers take part

Tourists of yester-year were generally **passive tourists**. They looked at the scenery, toured the static displays in museums and country houses and ate food as similar as possible to that of their own culture.

Fifty years ago most tourists wanted a holiday in which they could partake in basic physical activities such as swimming in the sea, walking in upland areas or skiing in mountains. Today we see a wide range of more active physical pursuits in addition to the earlier simple activities, e.g. white water rafting, snowboarding, pony trekking as well as holidays catering for specialist interests e.g. golf, cooking, adventures.

Most of these activities are found where the physical conditions are most suitable. However, there has also been the development of the Theme Park in which the thrills of 'white-knuckle' rides have been created by capital expenditure and these tourist venues are amongst the most popular in the world.

Weather and Climate

Each of the main climatic regions of the world has the potential to develop tourism based on the particular advantage that its temperature and rainfall bring. Florida with its sub-tropical climate is known as the sunshine state and attracts tourists all year round because of its warmth. The South of France also offers all-year-round warmth which is appealing to people from Northern Europe. Places with mountain climates where temperatures regularly drop below freezing attract winter sports enthusiasts.

Some climatic zones are less attractive for visitors, although even Lapland is marketed as the home of Father Christmas and has become increasingly popular!

CASE STUDY

St. Tropez and the French Riviera

The stretch of coastline between St. Tropez and Menton, on the Mediterranean coast of France has been famous since the mid-nineteenth century as the most fashionable resort area of all Europe. Its natural advantages are:

- a warm temperate climate with mild winters and hot dry summers;
- superb coastal scenery consisting of sheer cliffs formed by tertiary ridges of the pre-Alps, separated by inlets, coves and bays;
- a relatively calm sea with a low tidal range, creating ideal boating conditions;
- good road, rail and air links with the rest of Europe.

Along this 150 km stretch of coastline, serviced by the superbly engineered corniche autoroutes, are found some of the most expensive properties in Europe – built on cliff-top promontories with private beaches below. There are also thousands of hotels, apartments, caravan and camp sites catering, on a descending scale, for those whose wealth is less evident.

St. Tropez is a place where all these extremes meet and mix. It was principally a fishing village favoured by writers and impressionist artists who were attracted by the quality of the light and the colours of the landscape but became world famous in 1956 when Brigitte Bardot filmed *And God Created Woman* there. Today its resident population of 7000 increases tenfold in summer as both young and old, famed and wannabes come to see and be seen. There are innumerable bars, restaurants, boutiques, excellent beaches and lively night-clubs.

The resort retains its charm although it is swamped with people; like the rest of the Riviera, the crowds create the ambience.

Seasonal variations

The hot dry weather of Mediterranean summers forms the main reason for the popularity of France, Spain, Italy, Greece and Turkey during those months. Most regions with sufficient snowfall to enable skiing to take place have installed the basic equipment of lifts and runs. This applies to unlikely countries such as Morocco and Cyprus. Heavy rainfall during the monsoon season in the tropics deters travellers, particularly those looking for beach holidays.

International and domestic political situations

Destinations which are perceived to be dangerous or offer life-threatening situations are generally unpopular with all but a tiny minority of potential visitors. War in former Yugoslavia halted the expanding tourist industry of that country. The break-up of the Communist bloc has given large numbers of Eastern Europeans the freedom to travel, particularly into neighbouring Western Europe.

CASE STUDY

Egypt

Egypt has had a tourist industry ever since the nineteenth century. Its location on the southern shores of the Mediterranean meant that it could be reached fairly easily from Italy and France and once the Suez Canal had been opened in 1869 it became a port of call for thousands of people travelling between India, Australia and the Far East by boat. In addition its wealth of archaeology and architecture led it to become a fashionable destination. Educated wealthy Europeans and Americans travelled to Egypt to visit the tombs in the Valley of the Kings and to marvel at the treasures in Thebes and Luxor. Today these visits are often combined with a week's cruise on the River Nile. Together these activities bring in millions of tourists annually whose contribution to Egypt's balance of payments is vital.

In the past 20 years Egypt has been affected by the growth of militant fundamental Islam, like many other Middle Eastern states. These extremists wish to bring down the Egyptian government and have targeted tourists as a way of obtaining world publicity and destabilising the country. The figures for tourist arrivals for part of the 1990s show the success of their tactics (see Figure 1.19).

It may take the tourist industry many years to recover from the 1997 incident. Rock-bottom tour prices, special offers and other inducements are often insufficient to persuade visitors to return.

Year	No. of arrivals (millions)	% Change	Incidents
1992	3.2	—	Terrorist attacks on tourists
1993	2.5	−22	Government arrested and imprisoned known extremists
1994	2.58	+3	
1995	3.1	+21	
1996	3.8	+22.5	
1997	n/a		Terrorist attacks on tourist bus at Luxor. Many killed.

(Source: WTO)

FIGURE 1.19 The effect of terrorism on tourism in Egypt

The economic situation in origin and destination countries

Less economically developed countries of the world have grown as tourist destinations in the last 35 years and the revenue brought into the country forms a vital element of GDP.

Very few people from LEDCs travel abroad as tourists, although many Muslims from Asia and Africa may endeavour to make a pilgrimage to Mecca once in their lifetime.

The more economically developed countries provide both origin and destination for most of the world's tourist traffic. The top 15 destinations shown in Figure 1.18 are mainly in this group. Figure 1.20 gives the first 15 countries by tourist expenditure. This reveals that over 50 per cent of world tourist expenditure is generated from just five countries.

FIGURE 1.20 Expenditure on tourism 1995

Rank	Country	US$ million	% Share of world total
1	Germany	47 304	14.7
2	United States	44 825	13.9
3	Japan	35 322	11.0
4	United Kingdom	24 625	7.6
5	France	16 038	5.0
6	Italy	12 366	3.8
7	Netherlands	11 050	3.4
8	Austria	9500	2.9
9	Canada	9484	2.9
10	Taiwan	8595	2.7
11	Belgium	7995	2.5
12	Switzerland	6543	2.0
13	South Korea	5919	1.8
14	Sweden	5109	1.6
15	Mexico	4950	1.5

(Source: WTO)

The Monetary Exchange Rate

If a country's currency is strong on the world monetary markets, it becomes less desirable as a destination (foreign visitors have to spend more of their own currency unit) and more desirable as an origin. The opposite is obviously also true – a country with a weak currency is attractive to foreign visitors but deters its inhabitants from travelling elsewhere. During the 1980s, the Yen was one of the world's strongest currencies and Japanese tourists travelled all over the world. During the late 1990s the Yen did not hold such a favourable position and the Japanese tended to remain closer to home for their holidays.

The price of tourist services

Some destinations have become popular because they provide 'cheap holidays'. Spain, during the 1970s and 1980s was an example of this, but rising prices led to it being overtaken by Greece and Turkey as sources of cheap holidays in the 1990s.

The staging of special, short duration attractions and events

International sporting occasions such as the Olympic Games and soccer's World Cup attract large numbers of visitors for the duration of the event. The Passion Play at Oberammergau (Austria) only occurs every 10 years but is a major attraction at that time. Cultural festivals are held in many

FIGURE 1.21 First order international tourist flows (after Pearce)

cities around the world e.g. Mardi Gras (New Orleans), the Wagner festival at Bayreuth (Germany) and the Chinese New Year (Hong Kong).

Some of these factors are relatively stable and unchanging e.g. the weather, and can be thought of as responsible for the underlying patterns, whilst others may vary or even fluctuate within a relatively short period of time e.g. monetary exchange rate. These tend to have a greater short term effect but do little to alter the underlying pattern of flows.

Figure 1.21 shows a map of first order international tourist flows. Arrows are drawn to each destination from the country which provides its largest number of visitors. Comparable data was only available for 134 destinations. It reveals the dominance of the industrial countries of North West Europe, North America and Japan as tourist origins and the importance of southern Europe and the West Indies as destinations.

STUDENT ACTIVITY 1.4

1 a) Use the data in Figure 1.18 for the market share percentage of world totals for 1997 to draw a choropleth map with no more than five classes.
b) Choose one country from each choropleth class you created for the map. For each of these countries, comment on the factors (resources, activities, weather, political and economic stability) which have contributed to their ranking.
2 Represent the data shown in Figure 1.20 in the form of a graph. Comment upon the global distribution shown by the data.
3 Select one country from the list of major spenders in Figure 1.20. Use Figure 1.21 to list the favourite destinations of people from that country. What aspects of the destinations are likely to be attractive?
4 Choose a special event at any scale (local, regional, national or international) and research what impact it has had or is likely to have on the area you have chosen?

EXAMINATION QUESTIONS

1 Explain how and why the demand and supply of tourist facilities has changed in the last 50 years.
(50)
ULEAC

2 Briefly describe two ways in which flows in world tourism have changed in the last 20 years.
(4)
Cambridge

3 Figure 1.E.3 shows the growth of international tourism since 1950. Explain why such rapid growth has taken place, and identify specific areas in the world which have been particularly influenced by this growth.
(9)
Cambridge

FIGURE 1.E.3 The growth of international tourism

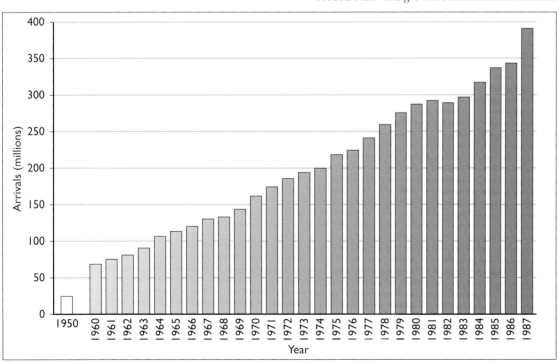

(Source: Cambridge Exam Board)

4 a) What are the main reasons for the rapid expansion in world tourism in the last 30 years? **(7)**

b) Figure 1.E.4 shows foreign holiday destinations of the British between 1970 and 1990.
(i) Identify the main changes which have taken place in foreign holiday destinations over the period 1970–1990. **(5)**
(ii) What further information would you need in order to obtain a more detailed picture of changes in foreign holiday destinations of the British over this period? **(3)**

c) Consider the advantages of tourism to the economic and social development of countries in the developing world. **(10)**

Cambridge

5 a) Show, with examples, how and why Britain's seaside resorts developed in the nineteenth and early twentieth century. **(15)**

b) Outline some of the problems created by this rapid development. **(10)**

O & C

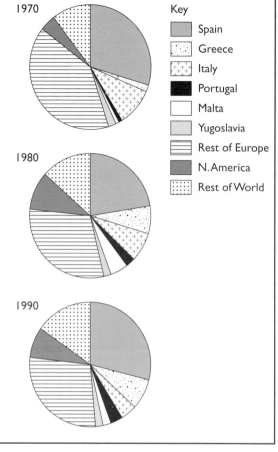

FIGURE 1.E.4 Holiday destinations of the British

Brighton

'... Brighthelmstone was only a small obscure village, occupied by fishermen, till silken Folly and bloated Disease, under the auspices of a Dr Russell, deemed it necessary to crowd one shore and fill the inhabitants with contempt for their visitors.'

Peregrine Phillips, 1780

Early beginnings

There has been a settlement in the vicinity of Brighton since the New Stone Age. For much of its history it has relied upon the sea as a means of economic survival. Until the mid-seventeenth century this was as a fishing town with a little piracy and smuggling on the side. However, in 1750, a Dr Richard Russell – medical practitioner of Lewes, East Sussex – published a book about the use of sea water as a treatment for afflictions of the glands. In it he recommended drinking and bathing in sea water as a cure for a variety of common ailments. The English translation (from Latin) three years later allowed the idea to become widely known and it suddenly became the 'thing to do' for the world of fashion. Brighthelmstone was the nearest seaside place to Lewes (a mere 15 km away) and had two inns to cater for the patients of Dr Russell. The fishing village grew steadily to accommodate the increased number of visitors to what was in effect the first seaside resort.

Royal Patronage

Brighton's future was assured in 1783 when the Prince Regent (later George IV) bought some land on the edge of the town and commissioned the building of a 'marine pavilion' for his use. The fashionable world soon followed, travelling the 80 km from London not only to enjoy the health benefits, but to display their wealth and privilege. Brighton quickly acquired a reputation for raffish behaviour and illicit pleasures and for 40 years it was virtually the second capital of England.

The Royal Pavilion (see Figure B1.1), as it is now known, was completed in 1821. It was considered a masterpiece of bad taste by many of the Prince's contemporaries with its Indian exterior and Chinese decor inside, but has remained a strong attraction for visitors.

The town grew rapidly – as illustrated by the number of houses recorded in the period:

YEAR	1770	1800	1821	1831
Houses	650	1300	4000	7000

TOURISM – A GROWTH INDUSTRY

Saved by the train

Victoria ascended the throne in 1837 and soon decided that she was not amused by Brighton. The town which mainly existed to provide leisure for polite society was suddenly not to Royal tastes. The Royal Pavilion was no longer in use and was only saved from demolition in 1850 by the forerunners of the borough council, who managed to buy it from the Crown (for £54 000)

Fortunately for the town's economy, Brighton was one of the first towns to benefit from the railway. Cheap excursion fares were introduced on the London to Brighton line in 1844 and so the town suddenly became accessible to the capital's working classes. An estimated 250 000 visitors a year were visiting Brighton by 1861 and in 1862, on Easter Monday alone the trains carried 132 000 passengers into the town.

Purpose-built attractions for this influx of visitors included the West Pier, Palace Pier, an aquarium and Volk's electric railway, all of which were financed by private capital. Penny arcades, souvenir shops and places of entertainment multiplied in those areas close to the beach.

The tourist business continued to grow through the 1920s and 30s. The raffish reputation of the Regency period returned as Brighton became infamous as the place to go for a 'dirty weekend' – where people cheating on their marriage partners would meet their lovers for illicit relationships.

Post war

During the 1950s and 60s Brighton advertised itself as a provider of sea, sun (but not sand – the beach is mostly shingle), and a variety of amenities and entertainments but with the growth of cheap air travel and the birth of package holidays, the town could no longer compete in the mass tourist market. Conferences had been staged in Brighton as long ago as 1874 (The British Association for the Advancement of Science), using The Dome – a building originally designed as very palatial stables for the Prince Regent's horses. The Hotel Metropole was also used, but the Council decided to invest heavily in a new sea-front building (the Brighton Centre) to further attract and encourage conference business.

Local councils have been able to designate groups of buildings of architectural and historical interest as conservation areas since 1976. Development within these areas is strictly controlled. Brighton began with seven areas which has now increased to 27 and has approximately 1900 listed buildings. Having focused on the sea for so long, Brighton seems to be offering heritage, history and hedonism as an alternative for the future.

FIGURE B1.1 The Royal Pavilion, Brighton

Spain

Spain is one of the world's leading tourist countries, both in terms of number of visitors and economic resources generated. Almost 62 million people visited Spain in 1996, so it is perhaps surprising to learn that until the 1950s the annual number of foreign visitors was under quarter of a million.

Reasons for tourism growth

Tourist development was spontaneous and unplanned, but growth was fast. The causes of this rapid expansion are partly those responsible for tourism growth generally, but specific reasons include:

Sun: A favourable position gives the Mediterranean coastal areas average summer temperatures of 24°C, winter temperatures of 12°C and a summer drought.

Sand: 40 per cent of Spain's 6000 km coastline is beach, mainly composed of white sand. Surveys of visitors show that almost 80 per cent give sun and beach as their main reasons for visiting Spain.

Accessibility: Spain has become one of the main destinations of European mass tourism because it is

FIGURE S1.1 The Mediterranean coast of Spain

saw that tourism would bring considerable economic advantages and introduced a number of measures to encourage further expansion. There were no regional planning mechanisms in place and foreign investors found few obstacles to financing hotel and apartment construction projects. Owners of previously almost worthless coastal land were prepared to sell to the highest bidder. Throughout the 1970s and 80s high rise blocks were built close together to accommodate as many tourists as possible near the sea.

Most tourists staying in Spain don't stay in hotels, but tend to rent villas and apartments or stay in second homes or timeshare developments. These latter two types of accommodation have contributed greatly to Spain's continued popularity by increasing tourist loyalty.

Some of the areas began to lose their Spanish characteristics, serving English and German food and drink. Many tourist businesses, especially bars were taken over by expatriates.

The Crisis

The number of foreign visitors to Spain fell in 1989 for the first time and then again in 1990 (Figure S1.2). At the same time unemployment was running at 20 per cent and there was widespread industrial unrest. The Spanish tourism industry was perceived to be in crisis. Responsibility for the decline was partly due to the dependence on too few origin markets (e.g. the UK and Germany) and on 'sun and sand' holidays. This reliance had left Spain vulnerable to fluctuations in the economy of their main markets (e.g. the mortgage rate in the UK doubled in 1989–90, considerably reducing consumers' disposable income) and to changes in consumer tastes. There was an 80 per cent increase in overall tourist prices between 1983 and 1990, a consequence of Spain joining The European Union in 1986.

The tour operators had mainly catered for 'cheap two weeks on the beach' holidays, but now these could be provided at a lower cost in Turkey and Greece. The higher tourist prices were more suitable for a different market, but the product offered by the tour operators didn't match up.

a relatively short, fast journey by road or plane from the rest of the continent. Approximately 80 per cent of visitors to Spain are from other EU countries. Admittedly half of these are from neighbouring, France and Portugal and tend to be day trippers, but over half of those who stay overnight are from the UK and Germany.

Cost: Until the late 1980s prices in Spain were below the European average. This had two benefits – tour operators could offer economical packages and still make a significant profit, and tourists had greater spending power during their holiday.

The double attraction of sun and sand led to early growth along the Mediterranean coastline. It is still the most popular tourist destination today – roughly 85 per cent of hotel capacity is in only five of Spain's 17 regions. These popular regions are the Balearic and Canary islands, Catalonia, Valencia and Andalucia (see Figure S1.1).

The Spanish government of the 1960s quickly

FIGURE S1.2 Visitors to Spain

Year	UK & Eire visitors	Total visitors	
1985	5 035 050	43 235 363	
1986	6 670 895	47 388 793	
1987	7 826 165	50 544 874	
1988	7 941 865	54 178 150	
1989	7 578 448	54 057 562	
1990	6 285 013	52 035 508	
1991	6 322 342	53 494 964	
1992	6 723 439	55 330 716	
1993	7 721 786	57 263 351	
1994	9 420 000	61 400 000	
1995	8 805 315*	58 350 000	* UK figures only
1996	not available	61 785 000	

(Source: Spanish National Tourist Office)

The Gambia

'Although the smallest independent country in Africa, The Gambia has much to offer all year round, from its unsophisticated charm, miles of palm fringed beaches, spacious tropical gardens and nature reserves to quality hotels, interesting excursions and some of the friendliest people on earth.'

The Gambia – Published by the Ministry of Tourism and Culture, 1995

History

The British gained control of the mouth of the River Gambia in the late sixteenth century during the scramble for land along the west coast of Africa. They built forts at Fort James and Fort Bullen on either side of the estuary to protect their possession of 11 000 square km, and their traders from the French. For the next 300 years they ruled The Gambia as the country became known, until 1965 when it became independent. The people consisted of two main ethnic groups, the Mandinka – a tribe of subsistent farmers and the Fulani – nomadic pastoralists found widely throughout West Africa.

It was during this period that hundreds of thousands of Africans were forcibly captured by slave traders and transported across the Atlantic to the West Indies and America to work as slaves in the agricultural plantations there.

Economy

Traditionally, the economy has depended almost entirely on the cultivation of groundnuts and their export in the form of nuts, oil and cattle cake. Tourism is the most rapidly expanding sector of the economy and currently contributes 12 per cent of The Gambia's GNP. Each year the number of tourist arrivals increases (see Figure G1.2), but The Gambia is still one of the poorest countries in the world being ranked sixteenth lowest by UN statistics based on GNP.

Most visitors come to enjoy the sun, sea and sand of a tropical country and to experience the magic of a friendly, relatively stable African country. The English colonial heritage means that most visitors are from Britain, although there are strong links with North America and Scandinavia.

Climate

The climate of The Gambia is such that although temperatures are not excessively high, Banjul has a January temperature of 23°C and a July temperature of 27°C. The summer months from June to October receive over 1000 mm of rain, making this period unsuitable for tourism. The winter months are the most popular time to visit a tropical country – temperatures are warm and the dry season prevails. (see Figure G1.2).

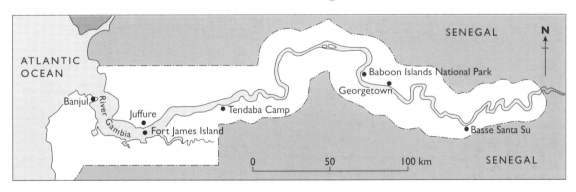

Figure G1.1 The Gambia

Month	1991/92	1992/93	1993/94	1994/95	1995/96	1996/97
July	1328	1426	4217	4142	3547	4774
August	1561	2110	3725	1938	2613	3695
September	1426	1597	4239	3398	3272	3233
October	2911	3684	5795	6565	3314	4526
November	8126	9736	9623	8252	9275	8838
December	11 098	9278	12 786	2988	7743	9415
January	10 157	8206	11 375	2543	8487	10 115
February	8776	6731	10 398	2547	9885	8644
March	10 444	8642	11 174	2549	9026	10 153
April	6608	5950	8073	3016	6884	7629
May	1821	3294	4558	2719	4415	5911
June	1515	3286	4034	2217	3636	5607
Total	65 771	63 940	89 997	42 919	72 098	82 540

Figure G1.2 Chartered tourist arrivals

(Source: The Gambia Tourist Office)

New Zealand

FIGURE NZ1.1 Map of New Zealand

The two major islands that make up New Zealand, North Island and South Island (see Figure NZ1.1) together comprise 268.112 square km. Altogether this is a land area slightly larger than that of the United Kingdom, but smaller than Japan or California. The islands are situated in the South Pacific Ocean between latitudes 34° and 47°; a location similar to that of the countries of Southern Europe, Spain and Italy. However, New Zealand is isolated in relation to other land masses. Australia is 400 km to the north west and Antarctica is 800 km to the south, otherwise it is entirely alone.

The north island of New Zealand has a sub tropical climate (Figure NZ1.2), while the south island is generally described as temperate. There is a strong maritime influence on temperatures, with less than 10°C variation between winter and summer. Rainfall is generally moderate with little seasonal variation, but the western mountainous areas receive far more precipitation due to their height than the eastern side of the country, which is drier and sunnier.

These variations result in widely differing types of landscape throughout the country. The mountainous western rim of South Island has an alpine appearance with its snow-capped peaks, glaciers and deep fjords; all clothed in dense forest (Figure NZ1.3).

FIGURE NZ1.2 New Zealand's climate

		Spring (Sept, Oct, Nov)	Summer (Dec, Jan, Feb)	Autumn (Mar, Apr, May)	Winter (Jun, Jul, Aug)
Bay of Islands	Temperature:	19°C/9°C	25°C/14°C	21°C/11°C	16°C/7°C
	Rain days:	11	7	11	16
Auckland	Temperature:	18°C/11°C	24°C/12°C	20°C/13°C	15°C/9°C
	Rain days:	12	8	11	15
Rotorua	Temperature:	17°C/7°C	24°C/12°C	18°C/9°C	13°C/4°C
	Rain days:	11	9	9	13
Wellington	Temperature:	12°C/6°C	20°C/13°C	17°C/11°C	12°C/6°C
	Rain days:	13	7	10	13
Christchurch	Temperature:	17°C/7°C	22°C/12°C	18°C/8°C	12°C/3°C
	Rain days:	7	7	7	7
Queenstown	Temperature:	16°C/9°C	22°C/10°C	16°C/6°C	10°C/1°C
	Rain days:	9	8	8	7

Daily maximum/minimum temperatures in celsius. (Source: New Zealand Tourist Office)

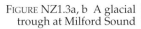

FIGURE NZ1.3a, b A glacial trough at Milford Sound

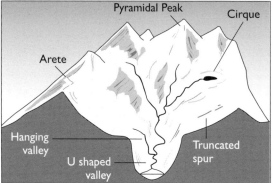

The exceptionally heavy rainfall is caused by the prevailing on-shore north-westerly winds which are forced to rise over the steep slopes of the Southern Alps. Over 5000 mm of orographic rain falls on the mountains, largely in the form of snow. Where the altitude is above 2500 m this remains throughout the year to accumulate as ice in the widespread gathering grounds. A huge amount of ice then spills outwards and down the valleys to create glaciers (see Figure NZ1.4) which radiate north, west and south of Mount Cook (the highest peak in the Southern Alps at 3764 m).

The whole area is visited by thousands of tourists. Few other places in the world have glaciers as close to sea level and the Franz Josef and Fox glaciers in particular are very accessible from the coast road. It is possible to walk to the snout of a glacier, or fly by sea-plane or helicopter and land on its surface. Alternatively, the area can be explored on foot, by four-wheel drive, horse-back or trail bike. The local towns and villages exist to cater for tourists. Earlier functions as whaling stations and logging ports are long since past. Hotels, motels and guest-houses provide accommodation for the non-stop stream of overnight visitors who come to see the spectacular scenery.

In the extreme south west of South Island the area of coast and mountains has been designated Fjordland National Park. Glaciers gouged great valleys down to the sea in this area, and were followed by post-glacial isostatic rises in sea level which caused the valleys to be flooded, creating many fjords (see Figure NZ1.5).

The eastern side is warmer, sunnier and less rainy. This area contains the wine growing regions and sheep stations that make up the mental picture of New Zealand for many people. North Island also consists largely of a farming landscape with dairying and orchards predominant on the east and west sides of the island respectively. However, it is the geothermal activity around Rotorua, Waikato and Taupo that forms the focus of tourist activity. Features such as volcanoes, hot springs, geysers and

FIGURE NZ1.4 Franz Josef glacier

FIGURE NZ1.5a, b (right and far right) Fjord formation at Milford Sound

FIGURE NZ1.6a, b (below and below right) Geothermal activity at Whakarewarewa

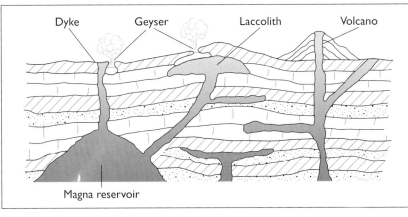

boiling mud occur when the crust of the earth is particularly thin or there is a zone of weakness that permits the heat from the mantle of the earth to be transmitted close to its outer rim (Figure NZ1.6).

Sometimes molten lava erupts at the surface to form volcanoes, as at Mount Ruapehu, while surface streams and underground water are heated to create thermal pools and hot springs like the Pohutu Geyser. Again it provides unforgettable scenic splendour for the thousands of tourists.

New Zealand is the home country of the Maori people, a race of Polynesian descent. They canoed from their legendary homeland of Hawaiki (in the central Pacific) almost 1000 years ago and settled in the islands they called Aotearoa (the land of the long white cloud). Then, during the great age of discovery in the seventeenth and eighteenth centuries, the islands were found by European sailors, initially from the Netherlands (hence Zealand, after the province of that name in South Holland). However, they were largely settled by migrants of British descent, who signed the Treaty of Waitangi with the Maori inhabitants in 1840. This gave European settlers the right to live in New Zealand, but provided protection for the Maori and their resources. Today there are 3.6 million New Zealanders, of whom 12 per cent are Maori. Their traditions, art forms and homelands have become part of the cultural heritage of the country and a significant element of its tourist appeal.

The Disney Concept

'Disneyland isn't just designed for children. When does a person stop becoming a child?'

Walt Disney

It was a film tycoon, Walt Disney, who first developed the idea of a theme park in the early 1950s. His initial plan was to create a place where families could enjoy a day out, and the first location was in California, on the southern edge of Los Angeles, close to the Disney film studios. It consisted of a central outdoor hub or plaza which gave access to four fantasy realms – Fantasyland, Adventureland, Tomorrowland and Frontierland. Here the visitors could experience for real the thrills and scares of roller-coaster rides as well as encounters with some renowned characters from literature (including fairytales and nursery rhymes), legend and history.

Within a week of opening, more than 1 million people had visited Disneyland, as it was called, and the concept of the Theme Park became part of the foundations of the tourist industry. By the late 1960s Walt Disney was looking for somewhere to open a second theme park. One possible site was in Tokyo, another was in Florida. Eventually the town of Orlando in Florida was selected. Its warm, sub-tropical climate meant that it could be open for business every day of the year, while the nearby freeways gave easy access to the site from the rest of Florida.

Disney and his associates mounted a secret campaign to purchase the huge area of land required without causing a massive rise in land values. So the citrus orchards of central Florida became a second Disneyland, known as The Magic Kingdom, in which the characters of Disney's many films could be encountered. Five km away the EPCOT centre (Experimental Prototype City of Tomorrow) focused on the ideas of science fiction for its fascination and thrills.

Today Orlando is the most popular place on earth. In addition to the Magic Kingdom other film companies and entertainment corporations have located their theme parks (for example Universal Studios, Sea World, Wet 'n' Wild) in the town, whose function now is to provide all the services such as hotels, hire cars and restaurants required by millions of tourists each year. Disney's initial idea to provide a 'magical little park' next to his film studios in California for the weekend recreation of his employees, has developed into a huge success story. Hundreds of imitation theme parks have sprung up all over the world. Disney created a completely new form of tourism.

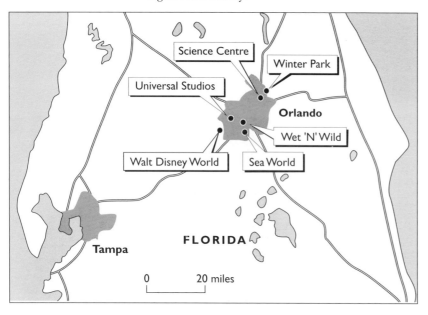

FIGURE D1.1 Theme park locations in Orlando, Florida

2
TOURISM AND SOCIETY

Key Ideas

- Increased demand for tourism may involve stress and change within communities
- Tourism poses questions about exploitation
- Transport network development has had to respond to the expanding global tourist industry
- The growth of tourism has important implications for the provision of accommodation
- The growth in incomes and leisure time has created a demand for the provision of increasingly sophisticated amenities
- The provision of amenities has significant impacts upon communities

What is society?

'Man seeketh in society comfort, use, and protection'

Francis Bacon The Advancement of Learning

Society is both the association with one's fellow citizens and the state or condition of living in association, company or intercourse with others of the same species (as defined in the Oxford English Dictionary). This is interpreted loosely here to allow an examination of the effect of tourism upon both the visitor and the hosts and also upon the built (man-made) environment in which the association between them takes place.

Social Issues

The interactions between visitor and host are complex and may involve stress and change for both people and communities that will increase as demand for tourist provision and amenities increase. At the very least, the presence of visitors means the hosts encounter someone from a different place. For countries in the developed world this can be a considerable shock. Spanish hoteliers generally classify each visiting nationality by common stereotype. The French are pushy and bad mannered, the Germans are stingy, the English are arrogant and the Italians are untrustworthy! In a less developed country the contrast may not only be a shock, but also a threat. If large numbers of tourists with a completely different system of values arrive, a long way from the constraints of their home environment, and then behave in a way that is offensive to the host nation, friction will occur. For example, it offends the Turkish people and is against their religious beliefs that women should appear 'topless' at any time. To do so is to risk arrest and imprisonment.

Along with offensive behaviour there will be problems of drunkenness, increased crime, drug taking and loutishness. These beset many of the larger resorts in Europe, North America and Australia. Many communities may well decide they were better off before the visitors came.

Gender Issues

There is inevitably a gender perspective to tourism because it is an activity involving both men and women, whether as participants or providers. Tourism also enables messages about gender to be transmitted, either deliberately or inadvertently. The holiday brochures which feature bikini clad girls to sell their location reinforce the stereotype of the female as a sex symbol (see Figure 2.1) while it is suggested that many heritage sites promote the concept of the male as a dominant aggressor. For people working in tourism, the reality is generally one of fulfilment of the traditional roles available to their sex. Most surveys of tourist related employment show that women are more likely to be engaged in unskilled, part-time, seasonal, low-paid work, while men are found in the professional, managerial, supervisory roles. Surveys in Looe, Cornwall, showed that most

FIGURE 2.1 Gender issues

hotel workers were young, married or divorced women, with one or more children, and happy to be able to work part-time hours as a chambermaid, waitress or receptionist to ease the problems of child minding. In Ireland, where similar patterns prevail in the hotel sector, it was also discovered that the 3200 approved bed and breakfast premises generated 7500 part-time jobs; 90 per cent being carried out by women. If one assumes there are thousands more bed and breakfast outlets than those officially listed, then a huge number of women are involved in low-paid household tasks.

In LEDCs the pattern is even more pronounced. Many women first become part of the work force of the country in a tourism-related role. It is an extension of their traditional role as homemaker, as occurs in both Cornwall and Ireland. They may also find themselves producing more food crops (this is generally their responsibility) or working additional hours to make the textile goods such as lace and embroidery for sale on craft stalls, as in Greece, Malta and North Africa.

The positive side to this picture is that women may be able to develop supervisory roles in association with these activities as hotel housekeepers, as manager of food-production co-operatives or as owners of shops selling craft goods. This progression can be seen in countries where tourism is long established.

In Western Samoa, women are not demeaned by tourism. The strong cultural traditions of the Polynesian tribes have given rise to *'faaSamoa'*, the Samoan way of doing things. Extended family groups share resources to promote the status of the family as a whole, and within this structure either a male or female could become chief. The person most suited to a particular role performs it. It is no accident that the tourist industry in Western Samoa is still based on family-owned hotels such as Aggies – 'indisputably the most well known hotel in the South Pacific', (*Aggie Grey: A Samoan Saga* by F Alailima 1988). Multi-national hotels such as Club Med are not to be found there.

Nevertheless there is continual concern lest tourism should damage *'faaSamoa'*. An advertisement for Polynesian Airlines depicting a scantily clad woman was withdrawn and replaced by one showing a Samoan male in traditional dress. The nightly entertainment featuring tribal dances is performed by both men and women wearing their traditional clothes. They will generally be hotel employees rather than groups 'brought in' for the occasion.

In many LEDCs it is felt that tourism has had a damaging effect on traditional values. Those areas within easy reach of North America, particularly

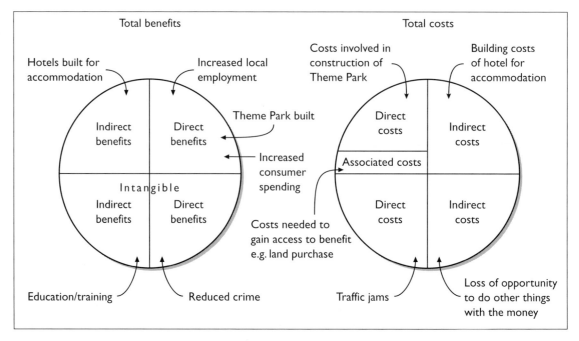

FIGURE 2.2 Costs and benefits associated with tourism (after Smith)

Barbados and the West Indies, are conscious that the arrival of large numbers of single men and women looking for casual sexual encounters has damaged the courtship and marital patterns of the local population. In addition, the visitors have introduced a hedonistic philosophy that is counter to traditional values. Some hotels in Barbados at one time would not accept bookings from single females, while the spread of AIDS throughout these holiday islands is an increasing problem.

In South East Asia the attitude towards tourists seeking sex has been very different. Thailand, The Philippines and Korea all have a reputation for readily available prostitution. This is presumably the motivation for many of the large numbers of single men who visit these countries from North America, Europe and Japan. The authorities operate systems of licensing and taxation to ensure standards of health and hygiene are met and to share the proceeds. It is only recently that voices of dissent have begun to make themselves heard. Such behaviour, it is claimed, is a form of imperialism, in which the exploitation of the local population is yet another form of colonialism.

Community Issues

Tourism as an industry has both costs and benefits to the area in to which it is introduced (see Figure 2.2.). Some are immediately obvious, some less apparent. For the tourist the quality of his or her holiday may be determined by the attitude of the host population that can range from warmth to hostility. Many political authorities, from governments to town councils, actively encourage tourism as a 'money-spinner'. A considerable number of people living in an area attractive to tourists gain financially, but this is no compensation when trying to find a parking space at the height of the season.

STUDENT ACTIVITY 2.1

1 Annotate a copy of the costs/benefits diagram (Figure 2.2) in relation to a recent tourist development known to you.
2 Discuss the extent to which gender is an issue in tourism.
3 Either for York or for a location you are familiar with, explain how tourism can become unwelcome and suggest measures that could be taken to reduce its impact.

CASE STUDY

The impact of tourism on York

The city of York has been inhabited for over 2000 years and evidence remains of its Roman, Viking, Mediaeval and ecclesiastical past. There is a wealth of famous buildings, a mediaeval street plan, city walls and various museums depicting conditions and artefacts of the past e.g. the Jorvik Castle and Railway museums. In the late 1970s, as manufacturing declined, there was a huge surge in the number of tourists who were attracted to Britain by the weak exchange rate. York's city council actively promoted the town to provide employment for those people who had lost their jobs in industry. The negative aspects of living in a popular tourist town were highlighted by the Yorkshire Evening Post on 31 May, 1994 (see Figure 2.3).

Creating a city for all to treasure

For people living in old and beautiful cities such as York, the tourist is as much reviled as he is welcomed. He is the person we love to hate but would hate to lose.

Residents of York, Cambridge, Chester, Bath, Canterbury, have all experienced the phenomenon of becoming a stranger in their own town as the summer hordes crowd in. At one point in the 1980s things became so pressured that badges were bought, stating 'I'm not a tourist, I live here' – until the tourists bought them, too.

York, with more and more visitors trying to cram inside its ancient city walls, enjoyed the spin-offs that the dollar and deutschmark brought, but became increasingly disenchanted with swarms of camera-toting foreigners blocking narrow streets and footways.

A report out today, from the English Historic Towns Forum, warns that some towns are now trying to discourage visitors as tensions grow over worsening congestion and pollution. Such is the crush that any pleasure in living in such historically rich cities is destroyed.

While not wanting to follow the example of Bath, where residents have turned hosepipes on open-top buses, or Cambridge, which considered holding a 'stay-away day', York must seize on the report's message that Government grants are needed to help manage the tourist influx.

One example of this is the development of Park and Ride for residents and tourists alike. With more sites planned, extra help would be extremely useful in laying on a proper welcome for visitors and enabling them to enjoy with us a city which all should treasure.

FIGURE 2.3 Negative aspects of living in a popular tourist town

Changes in types of tourists

The people who travel abroad now are very different from those who made foreign journeys in previous centuries. Exploits by travellers such as David Livingstone in Africa and Flinders in Australia are part of history. Foreign travel in the early part of the twentieth century had a similar sense of adventure and discovery. Now there are very few unexplored places left on earth to visit. The twenty-first century may well see the start of tourism in space or under the ocean. (In 1998 it became possible to visit the wreck of the Titanic at the bottom of the Atlantic Ocean, on a four day submarine trip.) This sort of adventurous travel only appeals to a small minority of the population. Most of us prefer not to boldly go where no one has gone before! A leading psychologist, Plog, recognises three different types of tourists according to their psychological characteristics. These are:

- Allocentric tourists – confident, adventurous, seeking challenges and new experiences
- Midcentric tourists – most of us, situated between the two extremes
- Psychocentric tourists – shy, home-loving, preferring familiar, safe surroundings

Each different type of tourist will want a different sort of holiday. It is the allocentric traveller who will venture alone to remote places, while the midcentric person buys a package tour to a foreign country and the psychocentric individual returns to the same location year after year.

Changes in travel patterns

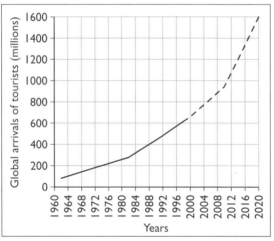

FIGURE 2.4 Growth in International tourist arrivals

(Source: WTO)

The number of global tourists is projected to continue to rise. Figure 2.4 shows the growth in the number of tourists' arrivals throughout the world until the year 2020. Figure 1.21 (page 17) revealed the importance of southern Europe and the West Indies as tourist destinations. However, if we consider the annual growth rate in visitors, rather than the actual number of visitors, another pattern emerges. Figure 2.5a shows that Europe has the slowest rate of growth whereas Oceania is growing much more quickly. Figure 2.5b shows the world's sub-regions with the highest growth rate in number of visitors. The fastest growing sub-region is Southern Africa, although it has less than 3 million visitors compared to the 50 million contributing to Central and Eastern Europe's lower growth rate.

FIGURE 2.5a International Tourist arrivals by world region

World region	Arrivals 1982	Arrivals 1992	Rate of average annual growth (%)	Share of arrivals (%) 1982	Share of arrivals (%) 1992
Africa	9 375 000	17 471 000	6.4	3.3	3.6
America	50 799 000	101 137 000	7.1	17.7	21.0
Asia	29 301 000	66 953 000	8.6	10.2	13.9
Europe	195 222 000	290 219 000	4.0	68.1	60.3
Oceania	2 083 000	5 785 000	10.8	0.7	1.2

(Source: WTO)

FIGURE 2.5b Sub-regional growth in International Tourist arrivals

World subregion	Arrivals 1988	Arrivals 1992	Rate of average annual growth (%)
Southern Africa	1 379 000	2 583 000	17.0
Central America	1 426 000	2 400 000	13.9
Micronesia (Oceania)	847 000	1 402 000	13.4
Southeast Asia	14 662 000	21 498 000	10.0
Central/East Europe	37 663 000	49 118 000	6.9

(Source: WTO)

TOURISM AND SOCIETY

The percentage share of arrivals by world regions is also changing. Europe's share is still three times greater than its nearest rival, but nevertheless it is declining. The number of tourists is projected to double between the end of the twentieth century and the year 2020. Where will they come from and where will they visit? The statistics suggest new areas will be developed or continue to develop at the expense of existing ones.

> **STUDENT ACTIVITY 2.2**
>
> 1 With reference to the range of holiday provision shown in Figure 1.17, (page 13), draw up a table to apply Plog's definitions of psychological types to each type of holiday.
> 2 a) Represent graphically the percentage share of arrivals shown in Figure 2.5a.
> b) Comment upon the growth shown in some regions in Figure 2.5b.

Changes in Communications

Chapter 1 covered the various forms of transport used by holiday makers in the last 100 years. It discussed their relative advantages and disadvantages and outlined their links to tourist growth. Holiday makers now have several transport options to choose from depending upon the holiday they wish to enjoy and their budget. For example a British holiday maker whose ultimate destination is Bruges (Belgium) could choose between:

- **coach travel** (cheap, slow, not door-to-door)
- **rail travel** (fast, expensive, not door-to-door)
- **Motorail** (fast, very expensive, door-to-door)
- **air travel** (very fast, relatively cheap, not door-to-door)
- **car** (moderately priced, slow, door-to-door, relatively stressful)

Figure 2.6 illustrates some of these differences.

FIGURE 2.6 Comparative prices of holiday travel

PRICES PER PERSON BY AIR

Holiday Ref: B71A

		1	2	3	4	5	6	7	†Mid Week	Extra Night	Single Room
HET PUTJE B4BH20	1 Jan–25 Mar	195	222	256	286	311	338	365	–	34	14
	26 Mar–31 Oct	199	229	256	294	321	348	376	–	35	13
	1 Nov–31 Dec	190	219	253	284	311	339	367	–	35	13
DE MARKIES B4BH30	1 Jan–25 Mar	207	232	263	292	316	342	368	–	33	13
	26 Mar–30 Jun	214	245	283	315	344	373	402	–3	36	15
	1 Jul–25 Jul	211	238	273	302	328	354	380	–	33	13
	26 Jul–31 Oct	214	245	283	315	344	373	402	–3	36	15
	1 Nov–31 Dec	199	227	257	286	311	336	362	–	33	13
OLYMPIA B4BH32	14 Feb–25 Mar	210	239	272	304	331	360	388	–	36	19
	25 Mar–31 Mar	210	239	272	304	331	360	388	–	36	19
	1 Apr–31 Oct	214	244	282	314	343	371	399	–	36	19
	1 Nov–31 Dec	204	233	267	298	326	354	383	–	36	19
NOVOTEL CENTRUM B4BH33	1 Jan–25 Mar	211	240	275	307	335	365	394	–	37	28
	26 Mar–31 Oct	224	263	311	352	389	427	465	–	45	23
	1 Nov–31 Dec	205	235	269	302	330	359	389	–	37	28
HANSA B4BH31	1 Jan–25 Mar	211	240	275	307	335	365	394	–	37	31
	26 Mar–31 Oct	221	258	302	341	376	412	447	–	42	29
	1 Nov–31 Dec	209	242	280	317	349	382	415	–	40	27
NAVARRA B4BH35	1 Jan–25 Mar	214	247	284	320	351	384	416	–	40	20
	26 Mar–31 Oct	224	263	311	352	389	427	465	–	45	22
	1 Nov–31 Dec	208	241	279	314	346	378	411	–	40	20
ROMANTIK PAND B4BH40	1 Jan–25 Mar	226	272	319	373	419	464	508	–	50	27
	26 Mar–15 Nov	232	284	340	389	437	484	531	–	53	28
	16 Nov–31 Dec	221	266	313	368	414	459	502	–	50	27

- PRICES BY AIR include return flights, return rail transfers, accommodation and breakfast.
- PRICES BY EUROSTAR include return Standard Class travel, accommodation and breakfast.
- † A midweek reduction applies for stays including Mon–Thu nights. • ALL holidays are subject to a minimum stay requirement – ask for details
- **Departures 9–12 Apr, 30 Apr–2 May, 21–23 May & 27–29 Aug add £15 per person** • *Please read the holiday Information and Booking Conditions on pages 162–163

PRICES PER PERSON BY EUROSTAR

Holiday Ref: B71E

		1	2	3	4	5	6	7	†Mid Week	Extra Night	Single Room
	1 Jan–25 Mar	139	169	199	233	258	285	312	–	34	14
	26 Mar–31 Oct	158	188	215	254	280	307	335	–	35	13
	1 Nov–31 Dec	148	178	212	244	270	298	326	–	35	13
	1 Jan–25 Mar	153	179	210	239	264	289	315	–	33	13
	26 Mar–30 Jun	173	204	242	274	303	333	362	–3	36	15
	1 Jul–25 Jul	169	197	232	261	287	313	339	–	33	13
	26 Jul–31 Oct	173	204	242	274	303	333	362	–3	36	15
	1 Nov–31 Dec	159	186	217	246	270	296	322	–	33	13
	14 Feb–25 Mar	156	185	219	251	279	307	336	–	36	19
	26 Mar–31 Mar	163	192	226	258	285	314	342	–	36	19
	1 Apr–31 Oct	172	203	242	274	302	331	360	–	36	19
	1 Nov–31 Dec	163	192	226	258	285	314	341	–	36	19
	1 Jan–25 Mar	157	187	222	255	283	312	342	–	37	28
	26 Mar–31 Oct	182	223	270	311	349	387	425	–	45	23
	1 Nov–31 Dec	164	194	228	261	289	319	348	–	37	28
	1 Jan–25 Mar	157	187	222	255	283	312	342	–	37	31
	26 Mar–31 Oct	180	217	262	299	336	371	407	–	42	29
	1 Nov–31 Dec	167	199	240	276	308	341	374	–	40	27
	1 Jan–25 Mar	160	194	232	267	299	331	364	–	40	20
	26 Mar–31 Oct	182	223	270	311	349	387	425	–	45	22
	1 Nov–31 Dec	167	199	238	274	305	338	370	–	40	20
	1 Jan–25 Mar	173	219	266	321	367	412	455	–	50	27
	26 Mar–15 Nov	191	243	299	348	397	444	490	–	53	28
	16 Nov–31 Dec	179	225	272	327	373	418	462	–	50	27

PRICES PER PERSON BY OWN CAR

Holiday Ref: B71F

	Number of Nights	1	2	3	4	5	6	7
HET PUTJE B4BH20	1 Jan–25 Mar	65	91	117	143	169	197	226
	26 Mar–30 Apr	89	116	143	170	197	227	256
	1 May–9 Jul	95	122	149	176	203	233	262
	10 Jul–31 Aug	99	128	155	182	209	239	268
	1 Sep–31 Oct	95	122	149	176	203	233	262
	1 Nov–31 Dec	66	93	120	147	174	204	233
OLYMPIA B4BH32	14 Feb–31 Mar	69	97	126	154	182	213	244
	1 Apr–30 Apr	92	120	149	177	205	236	267
	1 May–9 Jul	98	126	155	183	211	242	273
	10 Jul–31 Aug	104	132	161	189	217	248	279
	1 Sep–31 Oct	98	126	155	183	211	242	273
	1 Nov–31 Dec	69	97	126	154	182	213	244
HANSA B4BH31	1 Jan–25 Mar	70	99	128	157	186	218	250
	26 Mar–30 Apr	99	134	169	204	239	276	314
	1 May–9 Jul	105	140	175	210	245	282	320
	10 Jul–31 Aug	111	146	181	216	251	288	326
	1 Sep–31 Oct	105	140	175	210	245	282	320
	1 Nov–31 Dec	74	106	139	172	205	240	276
NOVOTEL CENTRUM B4BH33	1 Jan–25 Mar	70	99	128	157	186	218	250
	26 Mar–30 Apr	99	139	177	214	252	292	332
	1 May–9 Jul	107	145	183	220	258	298	338
	10 Jul–31 Aug	113	151	189	226	264	304	344
	1 Sep–31 Oct	107	145	183	220	258	298	338
	1 Nov–31 Dec	70	99	128	157	186	218	250

- Prices are based on 2 adults sharing and include a car ferry/Le Shuttle crossing, accommodation & breakfast.
- Additional adults travelling in the same car pay the full price less £20.
- Less than 2 adults, supplement of £20 applies.
- Single room & extra night prices are shown on page 31.
- **Please read Holiday Information & Booking Conditions on p. 162–163.**
- Departures 9–12 Apr, 30 Apr–2 May, 21–23 May ¶ 27–29 Aug add £15 per person.

The impact of these choices has far-reaching effects on people who live in the vicinity of major transport nodes, both national and international. The nightmare conditions of French roads during the last weekend in July (when their national summer holidays start) make it well known as a time for other nationalities to avoid travelling within France.

For people within earshot of an airport particularly one servicing holiday destinations, the constant arrival and take-off of planes is a major disturbance, although the increased employment opportunities offered may compensate. Nonetheless, these transport nodes are essential if a country is to develop its tourist industry.

Transport networks

Figure 2.7 shows the increasing numbers of tourists who have arrived at British air and sea terminals in the last 10 years. Their movements after arrival depend upon the range of transport links provided by national and local government and by private enterprise.

FIGURE 2.7 Changes in numbers of arrivals at British Air and Sea Ports

	1986	1988	1990	1992	1994	1996
	(Passengers in millions)					
Airport						
Gatwick	16.3	20.7	21.0	19.8	21.0	24.1
Heathrow	31.3	37.5	42.6	45.0	51.4	55.7
Luton	2.0	2.8	2.7	1.9	1.8	2.4
Stansted	0.5	1.0	1.2	2.3	3.3	4.8
Manchester	7.5	9.5	10.1	11.7	14.3	14.5
Glasgow	3.1	3.6	4.3	4.7	5.5	5.5
Source: CAA (UK Airports)						
Accompanied cars by ship (thousand vehicles)						
North Sea	572	605	650	734	632	547
Dover	1735	1583	2116	2606	3446	3150
English Channel	600	776	1140	1295	1436	1292
Eire & N. Ireland	631	717	915	1010	1164	1229
All other routes	306	321	399	387	431	454
Hovercraft services	218	262	230	109	181	233
Accompanied buses and coaches						
by ship and hovercraft	182	173	202	216	235	223

(Source: Transport statistics of Great Britain 1997)

FIGURE 2.8 Transport links from Gatwick

CASE STUDY

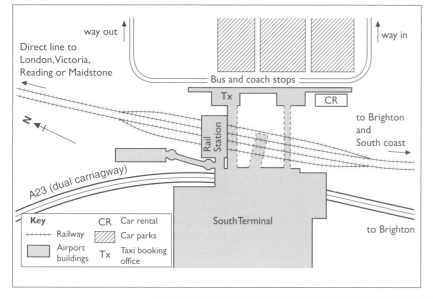

Gatwick Airport

Gatwick airport is the entry point for many overseas visitors to Great Britain. Figure 2.8 shows the range of transport links now available once a passenger has cleared customs.

The choice of transport type will depend largely upon the ultimate destination of individual passengers. A large number of arrivals are going to London either to remain there or to travel onwards. These people have a choice between train (from the airport station), coach link (from the coach concourse), taxi (from outside the airport buildings) or private car (which has to use airport parking facilities).

National governments have a responsibility to make sure these networks operate efficiently because poor service, time-tabling problems and/or long delays seriously affect a traveller's enjoyment of a trip.

Communication Networks

Many people use brochures or audio-visual presentations to choose their holiday destination. Their selected holiday may have several different components, e.g. airline seats, hotel beds, car rental, but these can all be booked through one travel agent or organisation. Individual travel agents do not physically stock these components until they are sold, but simply communicate and process the information about their availability. Customers' payments are also transferred as information in the form of debits and credits through electronic payment systems. Together this means that the tourism industry generates, gathers and processes a tremendous amount of information on a daily basis. Not surprisingly, information technology has been adopted in all areas of the tourism industry. Figure 2.9 shows the range of technology available.

customers to take cheaper holidays. Travel agencies can simultaneously search the airlines' data bases for the cheapest possible fare. They are also immediately aware of 'last-minute' bargains where airlines have spare seats available at big discounts in order to fill a plane.

It can be speculated that with the continued rapid spread of technology, future holiday makers will be able to make their own arrangements from their personal computer without the help of a travel agent or tour operator!

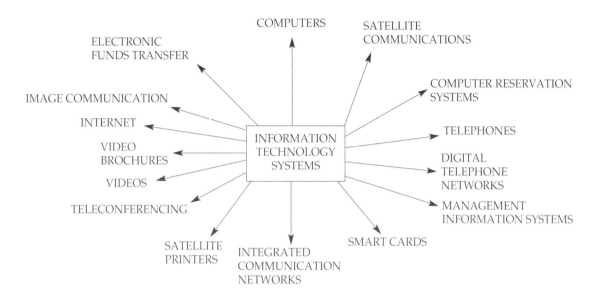

FIGURE 2.9 System of Information Technologies in Tourism (after Poon)

Computer reservation systems

Of all the various technologies available, the one that has had the greatest impact is computer reservation systems (CRS), particularly in the United States where over 96 per cent of all travel agents use computer reservation terminals. Their advantage is in increasing the efficiency with which travel agents can search for information about airline seats available, confirm bookings and reduce the airlines' costs for reservations and bookings. A survey of travel agents reported that installing a CRS raised productivity by 42 per cent and as a result the costs of airline reservation and ticketing fell by as much as 80 per cent.

Computer reservation systems have allowed

STUDENT ACTIVITY 2.3

1 Using the information given in Figure 2.6 (travel to Bruges), select one of the following groups of people and choose their most suitable mode of travel. Justify your choice.
– A family group with children under 10 years old
– Two OAPs
– A group of university students

2 a) Produce a flow line map to show arrivals at British air and sea ports in 1996 (from Fig. 2.7).
b.) Comment on the patterns shown by your flow line map.

3 Consider the range of options open to travellers landing at Gatwick who wish to:
a) go to London to continue their holiday
b) return to their home in Glasgow
c) stay overnight whilst in transit.

4 a) Use the internet, teletext or similar electronic media to find out about a possible holiday location.
b) Compare your findings with information offered in a brochure from a travel agent.

Accommodation

Figure 2.10 Banff Springs Hotel

Figure 2.11 Typical seaside accommodation

Tourists take themselves with them when they go away, so they have expectations of their holiday that will have been shaped by their home environment and life-style. They will expect accommodation that is at least the equivalent of what they are used to at home, unless they have decided to sacrifice standards for cheapness by camping or by staying in a caravan or cheap hostels.

Alternately, they may have decided to pay more for additional facilities such as swimming pools and other leisure facilities. Their choice of holiday accommodation reflects many psychological aspects of themselves. A package tour holiday in a chain hotel is extremely impersonal. It has been described as a 'bubble experience', in that living conditions will be similar to those of home in terms of hygiene, food and comfort. Opting for self-catering or bed and breakfast accommodation reflects an underlying need to recreate 'home from home'. Very few of us are 'total immersers' who want to live as the locals do!

The role of planners and tourist boards is to decide what sort of provision is needed in a particular town or area.

The impact of tourist accommodation

The first thing a tourist requires when he or she reaches the destination is somewhere to stay the night. The provision of accommodation has therefore always been closely linked to the form of transport used by the traveller. Coaching inns were established along main roads hundreds of years ago, spaced out at a distance covered by a team of horses in a day's travel. The immense hotels built by the Victorians at railway termini, both in Britain and in other parts of the world at classic beauty spots, such as The Trossachs Hotel near Loch Katrine, Scotland and The Banff Springs Hotel in the Canadian Rockies, had hundreds of bedrooms to provide accommodation for a train-load of travellers and looked more like palaces than hotels (Figure 2.10).

Their place has been taken in the late twentieth century by the airport hotel, often owned by an international chain such as Marriot or Sheraton, which can sleep a plane-load of people. In the meantime, a wide range of accommodation has evolved to meet every conceivable overnight need throughout the world. Figure 2.12 shows the various categories of accommodation available. This can be further classified into the number of rooms, types of services offered etc., and then graded according to the quality of provision by stars, crowns, rosettes or other symbols. Accommodation described as 'five-star' is known immediately to be immaculately decorated and furnished, with constant service for every need, and provision of a wide range of facilities – at a price!

TOURISM AND SOCIETY 35

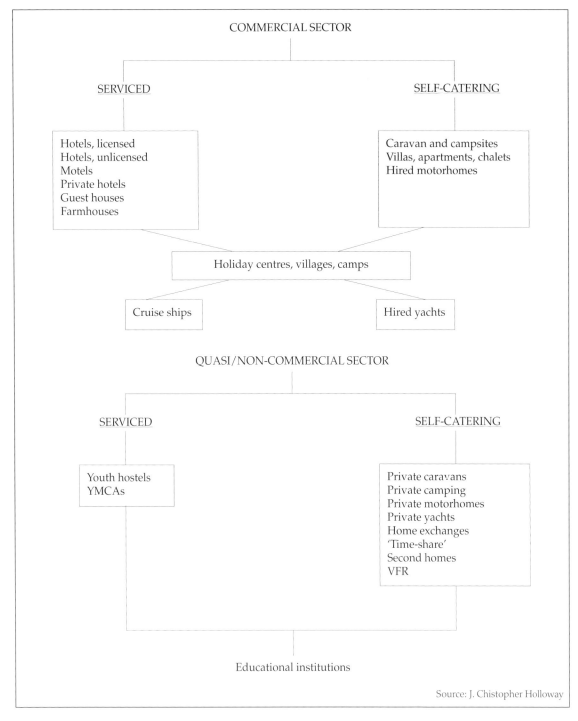

Figure 2.12 Types of accommodation available to tourists

Source: J. Chistopher Holloway

The English resorts of the early twentieth century such as Brighton, Bridlington or Broadstairs grew in response to the practice of going to the seaside with the family for a week's holiday from work. A variety of types of accommodation were built, ranging from large hotels to small boarding houses. Premises were splendid or shabby, licensed or unlicensed and catered for hundreds of thousands of visitors. They line the sea front and surrounding streets of most of our coastal resorts (Figure 2.11) and many have proved difficult to maintain in the later years of the century. Some have become flats or old peoples' homes. Others have managed to survive as hotels by creative marketing and extensive improvements of their facilities.

One of the most noticeable trends has been for forms of self-catering accommodation, whether in privately rented villas and apartments or in caravans and campsites at the lower end of the price range. It is true to say that throughout Europe, North America and Australia much of the beautiful coastal landscape sought by holiday makers has been destroyed by the spread of high-rise flats and caravan-sites. In some countries, e.g. Australia, USA and the Caribbean, ownership has also included the

FIGURE 2.13 Beach apartments in Queensland

coastline so that the beach itself is no longer accessible other than to those people living in the immediate accommodation (see Figure 2.13).

Accommodation in cities

Large towns and cities have always had a range of hotels to accommodate business travellers, while the great heritage cities of Europe – York, Florence, Salzburg and Venice for example – welcomed those seeking to admire their buildings, paintings and music. Some of the hotels built in previous centuries are themselves now famous landmarks such as the George V in Paris, The Ritz in London and the Crillon in Venice. City Breaks are now very popular for short holidays of two or three nights' duration, made feasible by surplus capacity on both airlines and hotel rooms once the business community is back at home for the weekend. Together with thousands of tourists from the New World – America, Canada and New Zealand – who come to see their European heritage, there is now a counter-flow of Europeans travelling outside their continent to visit famous cities such as New York, Sydney and Cape Town. In addition there are great numbers of tourists from the countries of SE Asia whose rapid economic growth of the 1980s and 90s enabled them to travel widely. Most recently parties of tourists from Eastern Europe, who were unable to travel for decades under communist rule, are to be seen sightseeing in the major cities of Europe.

All these groups require accommodation at different levels of comfort and service, with the result that hotels and hostels are in great demand.

City Regeneration

Large cities have always been centres of educational excellence and have also housed the treasures collected by former citizens in museums and art galleries. Nowadays these centres try to entertain as well as inform their visitors, as in the Jorvik museum in York, which recreates Viking times. Many other cities have lost their prime function, particularly if their original growth was fuelled by the Industrial Revolution, when demand for coal and trade in products from factories nearby led to rapid expansion. Their centres have become run-down and derelict as factories, housing and most recently, retailing, have moved to the cheaper open spaces on the outskirts. City councils and development corporations have tried various methods to regenerate economic activity in these areas in order to improve their appearance, provide employment and revitalise the core areas of their regions. Most have opted for a multi-purpose scheme, generally focused on a physical or historic landmark, inclusive of tourist attractions. Government has assisted by legislation to enable compulsory purchase of land and the provision of grants to assist in reconstruction. In Europe large sums of money have been paid from the European Community regeneration funds. The Docklands corporation has transformed large areas alongside the Thames in London.

CASE STUDY

FIGURE 2.14 Albert Dock

The Albert Dock, Liverpool

In Liverpool the Merseyside Development Corporation, formed by an Act of Parliament in 1981, has been responsible for work on 865 acres of land in the Albert Dock (Figure 2.14).

Liverpool grew as a city whose main role was to trade with the ports of North America and West Africa and to act as an exit for the factory goods of North West England, in particular the cotton products of Lancashire. The warehouses to store these commodities were immense brick built structures, lining the banks of the River Mersey.

They were opened by Prince Albert in 1846 with a speech starting 'I have heard of the greatness of Liverpool, but the reality far surpasses the expectation'. By the 1970s these warehouses were empty and derelict as advances in marine technology meant that large cargo vessels could no longer dock in the Mersey at Liverpool. Today they have been adapted to a variety of new uses; the Liverpool Tate Gallery, housing, offices, shops, Granada television studios, The Mersey Maritime Museum and an auditorium featuring 'The Beatles Story'. Five million visitors flock to the area annually, generating new income for the city and making the Dock the most visited multi-user attraction outside London.

Spa towns

Places with natural springs of mineral water, hot water or, in some instances, sea water, were able to benefit from their curative properties to provide 'cures' or convalescence for people who were ill or needed a rest. These centres or 'Spa towns' as they came to be known, are found throughout Europe, Asia and North America. Hotels and boarding houses grew up in the vicinity of the treatment centre, while other amenities and facilities such as casinos, race tracks and concert halls were developed to provide entertainment for the patients when they were not sitting in steaming vapours or drinking litres of foul-tasting water! Bath and Harrogate (Figure 2.15) are the best examples of spa towns in England, while abroad places such as Baden-Baden and St. Etienne have a similar reputation.

Although the practice of 'taking the waters' has declined both in North America and in some parts of Europe, it still remains in France and Germany, where spa treatments are supported by the medical profession and subsidised by both private and public health insurance schemes. Today these towns also attract tourists because of the elegance of their architecture and the character of their public buildings, preserved from an earlier era. In North America, spa centres now also focus on fitness and healthy living. They are attracting a new, younger type of client. In other words they are re-inventing themselves for a more profitable future.

Second homes

In some parts of the country, particularly in rural and coastal areas, accommodation originally built for local people has been bought by those with sufficient money to own two homes. In the United Kingdom this has occurred widely along the south coast, the Cotswolds and in North Wales, where as many as 15 per cent of the houses in the Lleyn Peninsula are owned as second properties. This figure increases to 36 per cent in the parish of Llandegan. People who live in cities may decide to buy a country house or cottage for their own weekend and holiday use. They may take advantage of improvement grants for upgrading old property and derive income from additional lettings while intending eventually to retire to the home they have renovated. In some cases it is the family home in the country that will be kept on by the children after the parents have died. In Britain less than 1 per cent of property is owned in this way, but elsewhere a larger percentage is used as holiday accommodation. It is estimated that 7 per cent of housing in France falls into this category (some owned by English people) and 20 per cent of housing in Scandinavia.

FIGURE 2.15 The Pump Rooms, Bath

Generally it is regions of attractive landscape close to centres of large population that are sought after for second homes, but as travel times decrease, so remote areas may find that housing is being purchased by outsiders. This inevitably causes house prices to rise, to the disadvantage of the local people, whose income is often smaller. Other local services such as schools, village shops and bus services will be adversely affected. Resentment about the 'incomers' may reach the point of aggression as happened in North Wales during the 1970s, where holiday cottages were burned by local people. However, conversely, second home owners will be paying their local taxes but not making as many demands on local services as if they were living there throughout the year. They will contribute to the local economy through purchase of repair and renovation services and they will bring renewed vigour to the community if they settle there permanently. Their presence slows down the rate of rural depopulation that would otherwise occur.

Rural regeneration

The local population may not welcome individual newcomers in a rural area and is often very hostile to a large scale scheme. Research has shown that there is the highest chance of success when consultation has taken place and the need for renewal is accepted. Unsought fame, such as occurs when a TV series is based in a recognisable location, is most resented. Even if economic benefits occur there will be widespread social tension. Jobs that are created are often low-paid, seasonal and part-time. Traffic congestion and lack of parking space in country lanes and village streets will cause frustration and further tension. Some regions such as the Lake District and Peak District in England will alter beyond recognition as they become tourist honey-pots. Yet the inflow of money will have raised standards of living throughout the area and the appreciation of the visitors will rekindle pride in the place. The California State Tourist Office devised an initial questionnaire for use in rural communities who were considering using tourism as a means of economic diversification (see Figure 2.16). This consultation is a vital element of decision making if local residents are to appreciate the benefits of tourism rather than only suffer the costs.

FIGURE 2.16 A questionnaire for rural communities

Is Tourism For Us?
A PRELIMINARY QUESTIONNAIRE

Is your community dependent upon one industry?
☐ Yes
☐ No

Benefits
If so, perhaps tourism could diversify the economic base.

Costs
An expanded tourism industry could require additional infrastructure.

Are local businesses
☐ expanding?
☐ stable?
☐ declining?

Benefits
If 'stable' or 'declining', then tourism may provide a needed boost.

Costs
If 'declining', then improvements may need to be undertaken.

Is unemployment seasonal?
☐ No
☐ Yes
Which season(s)?

Benefits
If developed during the slack season, tourism may help.

Costs
Some residents may desire a slack season and resent the 'congestion' during the time they anticipate 'peace and quiet'.

Are the unemployed
☐ skilled?
☐ unskilled?

Benefits
If 'unskilled', they may benefit from an increased need for service workers with minimum skills. If 'skilled', you may have to creatively explore symbiotic connections (e.g. entrepreneurial activities).

Costs
If 'unskilled', then training may be both desirable and required through local schools or job-training agencies.

Is there an appropriate labour force available locally?
☐ Yes
☐ No

Benefits
If so, perhaps tourism could provide needed jobs.
Costs
If not, you may have to 'import' workers from nearby communities.

Have sales tax and other taxes, for instance, transient occupancy tax (TOT), revenues been
☐ increasing?
☐ declining?
☐ remaining stationary?

Benefits
Sales tax revenues increased by visitor purchases could relieve the burden on local residents. Increased TOT funds, for example, could pay for greater tourism promotion, thereby setting into motion an upward spiral.
Costs
Revenue is not immediately available; an initial expenditure for additional promotion must first be made.

Is the diversity of shops and stores
☐ inadequate?
☐ considerable?
☐ somewhere in between?

Benefits
If 'inadequate', expanded tourism activity may stimulate more diversity. If 'considerable', the diversity may be a drawing card for more visitors.
Costs
If 'inadequate', then a greater diversity may have to be encouraged, which is sometimes difficult and may require extensive economic development work.

Are your downtown areas and main entrances
☐ attractive?
☐ in need of clean up?
☐ in need of major restoration and repair?

Benefits
If 'attractive', you have a greater potential for attracting and holding visitors. If 'in need of clean up', community organizations might be mobilized and the results will boost community morale, as well as set the stage for increased tourism activity.
Costs
If 'in need of major restoration and repair', then funding will be required; however, this can be done gradually. After the initial stages, increased tax revenues may be used. Be sure to evaluate sign-posting, roadways, parking, public conveniences, waste disposal, and public safety.

On the issue of increased tourism activity, is your community in
☐ agreement?
☐ uncertain?
☐ at opposite poles?

Benefits
If 'in agreement', then you will have the support you need to market your area.
Costs
If 'uncertain' or 'at opposite poles', then you need to invest time for education and consensus building so residents and business people will be hospitable hosts to your visitors.

Are local cultural activities ☐ thriving? ☐ struggling? ☐ of top quality? ☐ silly?	**Benefits** If 'thriving' and 'of top quality', you may be able to quickly appeal to an expanded audience and thereby generate greater support from a broader base. You can then offer more and the community will benefit through expanded cultural horizons. **Costs** If 'struggling' and 'silly', you will need to upgrade. This will take time and may meet with resistance. There may be concern by local residents about 'sharing' space and activities with others.
Are there a lot of recreational activities? ☐ Many unique choices ☐ Yes, but you have to know where to look ☐ Not much	**Benefits** If you answered 'many unique choices', you are sitting on a gold mine. If you said 'yes, but you have to know where to look', you may have a gold mine once you conduct an inventory and arrange for display and distribution of information. **Costs** If you said 'not much', then you either have little potential to attract visitors, or you need to look at your community through the eyes of an outsider. A fresh perspective sometimes creates a new picture.

Rural Tourism Marketing. Rural Tourism Center, California State Tourism Office, USA, 1987.

This questionnaire is a preliminary step. Other important issues will also need to be considered, such as the physical environment of the community, the potential market and the presence of tourism in the area.

Purpose built resorts

In some cases a completely new settlement has been created in order to develop tourism further in a particular area of the country. Such a decision comes from central or regional government, who alone have the authority to give planning permission for such a venture. Both public and private money will be involved in the provision of access routes, infrastructure, accommodation and leisure attractions.

Communities do not always welcome the idea of new resorts springing up in their vicinity. The Sri Lankan government was forced to cancel a resort development in the Wanigasundera region of the country in 1994. The land wanted for a golf course, horse race track and a casino hotel would have destroyed coconut production over a wide area and local opinion was firmly against the idea. Similarly the Barbadian government decided not to permit the construction of more golf courses on the island of Barbados because of local fears that supplies of water, already stretched, would be inadequate.

CASE STUDY

The ski-resort. Courchevel, Savoie, France

Skiing originally developed as a winter activity for those people who lived in the mountainous areas of Europe or for those who were wealthy enough to be able to travel there in the nineteenth century. Consequently, amenities were few or non-existent. Early photographs show enthusiasts walking up mountain paths wearing seal skins over their skis so that they could enjoy the brief pleasure of skiing down. It was only in the 1950s and 60s with the advent of package tours that facilities and amenities for a mass market were installed.

The town of Courchevel in the Department of Savoie in Alpine France was one of the earliest purpose-built ski resorts (see Figure 2.17). Thanks to a far-sighted mayor and regional councillors with vision who wanted to broaden the economic base of Savoie, a site was chosen high in the Alps which offered superb skiing once it was opened (see Figure 2.18).

Maurice Michaud, the chief project engineer had this to say: 'Rather than look for a village where we could put ski-lifts, we decided to find an ideal place for skiing where we could put in a village'.

Buildings were to be clustered at the base of the ski-lifts, which fanned out into the surrounding hills and plateaux. These were deliberately sited at 1500, 1650 and 1850 metres to provide different grades of skiing and different styles of accommodation (see Figure 2.19).

Today, the resort is further extended, with additional villages at 1000 and 1300 metres, especially geared to families. The whole venture includes 600 km² of slopes with 200 ski-lifts of differing sorts (gondolas, cable cars and chair lifts) able to transport 68 000 skiers per hour to the upper slopes, from where they choose between 150 km of downhill runs or pistes. Twenty-two machines work throughout each night repacking the snow on the runs, while 479 cannons stand by to manufacture artificial snow. Theme parks provide variety for snow boarders and for fun skiing, and leisure centres provide alternative winter activities. The resort has 32 000 beds, 14 per cent in hotels, 29 per cent in self-catering apartments, 43 per cent in chalets and the remainder in private houses and hostels. In general, the upper resorts have the most expensive accommodation, the most exotic night life and the best transport facilities (they are serviced by a small airport at 2000 metres which is accessible to helicopters and executive ski-jets). The journey time from Paris is only five hours. Sixty-three per cent of visitors to this premier resort are French, 20 per cent come from Britain.

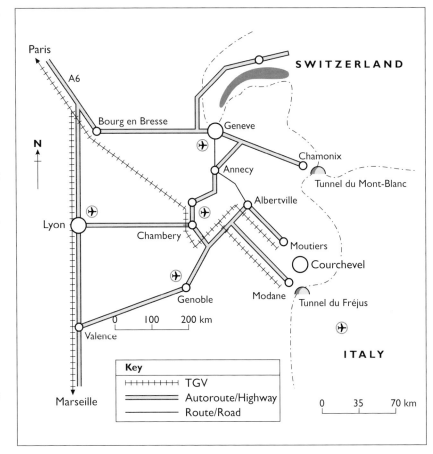

FIGURE 2.17 (above) The location of Courchevel

FIGURE 2.18 (left) The site for Courchevel

STUDENT ACTIVITY 2.4

1 Stidy the different styles of accommodation shown in Figures 2.10 to 2.15. To what extent do they reflect changing holiday tastes?
2 Write a letter to a newspaper about the number of houses in your town which are being bought as second homes. Refer to the impact that this is having on local society and jobs.
3 a.) Working in small groups, apply the rural regeneration questionnaire to your local area.
b.) Produce a summary report to be sent to the local development authority.
4 Using the case study material for Courchevel, and any other materials available to you, draw a simple sketch of an Alpine ski resort (real or imaginary) to summarise the costs and benefits which mountain tourism brings.

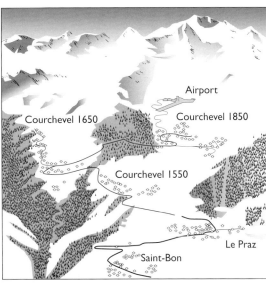

FIGURE 2.19 The lay-out of Courchevel

Amenities

In Chapter 1 we saw how the amount of free time increased during the second half of the twentieth century as advances in technology led to a shorter working week; while improvements in transport meant that both the time and cost needed to travel became less. Figure 2.20 shows the amount of free time enjoyed by the average European adult in 1995 by comparison with earlier years.

There has also been a change in social attitudes as people try to lead healthier lives and to exercise more regularly. Money was made available at every level from national government to private enterprise to build and equip the playing fields, leisure centres, concert halls and stadia that were demanded by the public. It is now possible to take part in almost any leisure pursuit you wish, even if neither climate nor physical feature were there in the first place. You can ski on artificial slopes, skate on ice rinks, abseil down climbing walls and practice scuba diving in the local swimming pool.

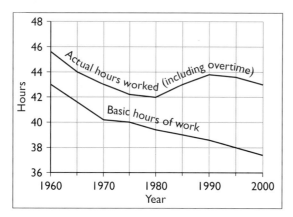
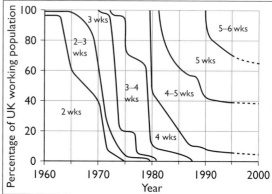

FIGURE 2.20a, b Changes in the working week and paid holidays in the UK

CASE STUDY

Leisure Activities in Sheffield

Figure 2.21 looks at the way in which the city council has built a train route to link the major leisure venues – the city with the city centre. The advertisement is targeted towards secondary school students who might be expected to travel across the city by tram to reach the various options illustrated here.

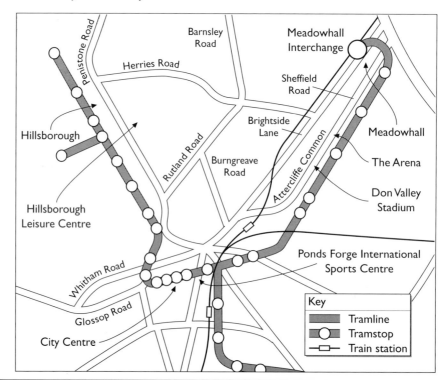

FIGURE 2.21 Public services to leisure amenities - Sheffield.

TOURISM AND SOCIETY

LEDCs also have a range of basic leisure facilities, particularly for sporting activities, and encourage their citizens to participate as fully as possible. Very often these countries will provide opportunities for visitors to indulge in leisure pursuits that would be impossible or illegal in their own countries (see Figure 2.22).

Private companies have also specialised in developing activity resorts where the holiday cost includes free use of various amenities. For example, wind surfing, water-skiing, scuba diving and yachting are offered to holiday makers at a Club Med resort.

The range of amenities and facilities found in tourist resorts has also increased considerably as visitors have arrived with much higher expectations of what they want to do on holiday than they had in the past. This is seen particularly in old towns and villages where the function of many buildings and the types of merchandise in shops will have changed.

FIGURE 2.22 Big game hunting

CASE STUDY

Alfriston, East Sussex

This was originally a farming village in the South Downs, a few miles from the sea. Today it is a tourist honeypot and many of the old buildings have different uses (Figure 2.23).

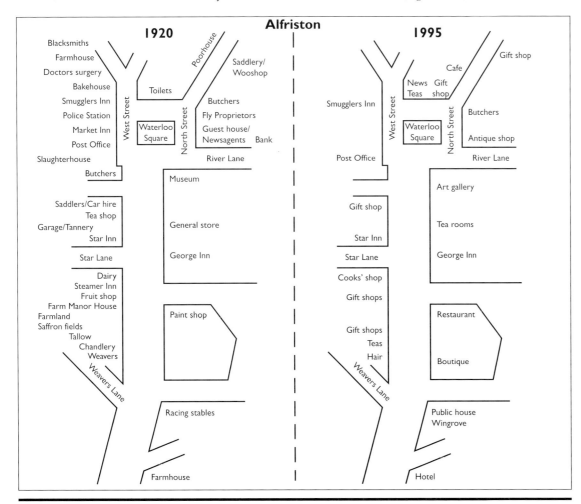

FIGURE 2.23 The changing functions of Alfriston, East Sussex

Classification of amenities

Figure 2.24 shows the classification of amenities as recognised by the World Tourist Organisation. In addition to the shops and businesses whose total income is from visitors, there are many services and amenities that are largely dependent on tourists but also provide a useful service to residents, e.g. taxis, restaurants, theatres and bars.

FIGURE 2.24 WTO classification of amenities

Recreation	Recreation and Sporting amenities
	Cultural activities
	Entertainments
Shopping	Souvenirs
	Duty Free
	Clothing & Footwear
	Luggage
	Tobacco Products
	Personal Care Products
	Other Goods
Other Services	Financial Services
	Travel items not included elsewhere
	Health/Medical
	Education or Training
	Other Services

Amenities at holiday resorts

The mix of activities, shops and entertainments available at tourist destinations will determine who visits and how successful their holiday is. There is evidence that clustering of similar amenities adds to the success of the location. Tourists are now sufficiently discerning and selective to choose a holiday destination that has the facilities they enjoy. British teenagers flock to Ibiza to enjoy the primary attractions of sun and sea and the secondary opportunities to sample the bars, clubs and discos which operate throughout the night. Blackpool provides less sun and colder sea but has a range of indoor attractions such as amusement arcades and fun fairs which stretch for miles. The electric lights associated with these entertainments have themselves become a draw for tourists in their own right. However, all these 'sights' tend to become less fashionable as time passes. Business growth and decline is cyclical and this applies to tourist investment. Unless there is constant introduction of new attractions, games and thrills, the tourists and visitors will stay away and go to the resort or theme park offering the latest rides and entertainments.

Theme Parks

The rapid growth of theme parks which offer a range of activities for an inclusive price has been a feature of tourist development over the past 35 years. This idea started over 70 years ago when Billy Butlin opened the first of many holiday camps in England. Visitors were accommodated in chalets and fed in mass sittings in a canteen. The attraction of these holidays was their low cost and inclusive entertainment. There are only five camps remaining open today – several of the early camps are now in use as prisons! The basic concept of providing activities along with accommodation is now widespread. For example, even small caravan sites will include a children's play area. At the other extreme, a Dutch company, 'Centre Parcs', have developed the idea of an all-inclusive holiday to the extent that there is no need to venture outside their climatically controlled domes while staying on one of their sites. Their popularity can be gauged from the fact that they report 90 per cent occupancy at an annual level. For those people who wish to have a totally inclusive day excursion, a number of theme parks have been built in the vicinity of major cities throughout Europe. Fantasialand near Koln has 10 million people within one hour's drive, as does Parc Asterix to the north of Paris and Thorpe Park, south west of London (Figure 2.25).

These parks are located close to major centres of population because they need to have thousands of visitors daily if they are to recoup their capital investment. There will be mixed reactions from residents in the vicinity of a proposed theme park as the 'Disney concept' recurring case study in this chapter shows.

Special interests and special events

As leisure time increases, so the importance of leisure activities grows. More money is spent in pursuit of skills, hobbies and interests. It is a natural extension of this desire to improve and learn more that accounts for the growth of special interest tourism. The range of holidays available that include specialised activities is enormous.

FIGURE 2.25 Thorpe Park theme park

Fifty years ago these holidays could be simply classified as: sea swimming in the summer; skiing in the winter; all year round hill walking and climbing for the hardy. Now you can learn to cook in France, go white water rafting on the Zambesi, search for rare insects and birds in Ecuador or camp on a glacier in Antarctica!

People combine their holidays with visits to sporting and cultural events. They go to a particular place at a time when it is celebrating a local festival, such as Carnival in Rio de Janeiro or tulip time in Holland. This sort of tourism has become of major importance for those responsible for marketing a particular place. Hotels will be fully booked during a classic horse race and many visitors may stay on afterwards.

Cities offering to host major sporting events such as the Olympics or the football World Cup do so because of the publicity, the trade, the employment and the image enhancement that will result. Major building programmes are usually necessary to provide the accommodation, stadia and transport infrastructure needed for thousands of competitors and visitors. This provides the opportunity to regenerate run-down inner city areas using finance raised privately through loans and publicly by extra taxes. However, it raises the question as to whether it is right to replace the decaying urban core of many major cities with leisure facilities, especially when those displaced by the venture are generally the poorest.

Often a national government will bid for an event and will then upgrade its facilities if successful. The town or city will still have the facilities afterwards and will then be in a position to host future activities, though payment may be a burden on taxes for years to come.

CASE STUDY

Olympics 2000 – Sydney, Australia

Sydney, the state capital of New South Wales, has won the nomination to host the 2000 AD Olympic Games by one vote in competition with Berlin, Beijing, Istanbul and Manchester. The cost of this bid was Aus$20 million. They were anxious to be successful in order to regain their place as Australia's number one tourist location, having lost the title to Queensland in recent years. The planners have emphasised that they will organise a 'green' Olympic Games. Their bid stated '… Sydney's Olympic Village design, prepared in collaboration with Greenpeace, foreshadows the sustainable city of the twenty-first century' (Sydney Olympic Bid 1993). Solar power will be used, along with water recycling, efficient public transport, electronic mail, multi-use tickets and minimal food packaging. The New South Wales state government estimate the cost to be in the region of Aus$1.7 billion, but the national government claim it will be Aus$2.3 billion because they include the money needed to build the sporting facilities in Homebush Bay as an Olympic expense. Regardless of who is more accurate, the amount of money being spent for the Olympics means that other urban regeneration projects throughout Australia will not be funded during this period. On a more positive note, the opportunity to pioneer environmental urban technology will enable Sydney to market the concepts and designs throughout the world and thus recoup some of the expenditure. Local residents will have the long term use of the facilities, everyone will enjoy being the centre of world media attention with the opportunity to attend events; Sydney will prosper!

The role of tourism in society

Figure 2.26 shows that the tourist industry is closely linked to both community life and development and to the maintenance of the physical and cultural environment. The opposition expressed by protest groups in Berlin to the proposed expenditure for the 2000 Olympic Games was one of the reasons why that city was unsuccessful in its bid. In Malaysia recently, a joint meeting of organisations representing service workers has highlighted the amount of tourism related to prostitution in SE Asia, especially that involving child sex, and has demanded that it be controlled. A growing awareness of other peoples' needs may be one of the characteristic features of tourism in the twenty-first century. It can be said that if everyone acts responsibly, then visitors, residents and workers in the tourist industry can all enjoy an improved quality of life.

The holiday locations of the early and mid-twentieth century have become the permanent home for millions throughout the world as people realise they are able to enjoy the benefits of climate and scenery all year, particularly in retirement. It seems that the cities of the MEDCs are also becoming centres for relaxation and leisure in addition to their cultural role. Perhaps the future is one of sustainable domesticity or of all year round, stay-at-home tourism?

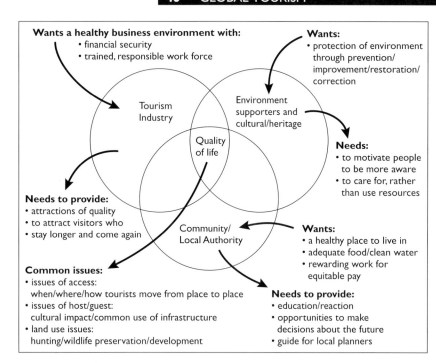

FIGURE 2.26 The links between tourism, the community and the environment

STUDENT ACTIVITY 2.5

1 Discuss the factors which have led to increased free time and paid holidays this century.
2 Investigate the range of activities available at your nearest leisure centre for different age groups. Is each group represented fairly? If not, suggest activities to correct the imbalance.
3 With reference to Figure 2.23 (Changing Building Functions in Alfriston), group the building functions, both in 1920 and 1995, that were:
a) provided for local residents
b) provided for tourists
c) provided for both.
Write a summary paragraph about the changes which have taken place.
4 Mark the location of Alton Towers, Staffordshire (the most visited theme park in 1996) on a map of Great Britain.
  Plot concentric circles centred on Alton Towers to represent 1 hour, 2 hours, 3 hours and 4 hours driving time to the theme park (assume 80 km can be covered in each hour).
  Mark the large urban areas included within the circles and find the size of their populations. Calculate what percentage of Great Britain's population (59 million) could:
a) Take a day trip to Alton Towers (2 hours drive away)
b) Spend a weekend at Alton Towers (4 hours drive away).

EXAMINATION QUESTIONS

1 What factors influence the balance between the costs and benefits of tourism's impact on human communities?
(50)
ULEAC

2 a) With a range of case studies, show how the demand for tourism can change human communities.
b) Suggest how you would develop a research project to establish how tourism benefits an area.
(25)
ULEAC

3 For any named area or areas you have studied, outline the impact of tourism on the local population, communities and cultures.
(9)
ULEAC

Brighton

The morphology or ground plan of coastal resorts is remarkably similar throughout the world. Its form reflects location and function. Location is dominated by the coast line, so morphology is linear or semi-circular. The function of coastal resorts is to provide for the requirements of holiday makers at the sea. Transport termini, accommodation and leisure amenities are arranged in the area behind the seafront, which in northern Europe forms a zone of 'promenade'. Figure B2.1 shows all the characteristics of the north European beach resort and can also be used as a model for Brighton.

We have seen in Chapter 1 how the royal patronage of Brighton gave shape to its early development, while the coming of the railway in the mid-nineteenth century brought the mass market to the town. Provision for the needs of thousands of day trippers led to the wide range of amenities and activities for which Brighton is still famous today. An extract from Kelly's Directory (1938) of land use along King's Road shows just how extensive this range was in those days, while the survey taken in 1998 shows the extent of change 60 years later (see Figure B2.2).

There are fewer buildings partly as a result of redevelopment and the construction of the Brighton Centre but their function has also changed. High class specialist shops have moved to other locations within the town and have been replaced by those providing cheap goods for day trippers along with food and entertainment.

Without continued re-invention (see Butler's model of tourism development in Chapter 3), holiday destinations go into decline. In the early 1970s when Brighton was faced with rapidly falling numbers of visitors due to the attractions of the Mediterranean coastal resorts, the Town Corporation carried out a three year survey, between 1972 and 1974, to ask visitors what they found most attractive about Brighton. To their surprise it was the miscellany of architectural styles and eclectic range of activities that was most appealing. In rank order visitors referred to:

- The Lanes (an area of pre-Georgian streets selling specialist articles from tiny shops)
- The Pavilion (the Prince Regent's palace)
- Regency squares (rows of terraced housing built around formal gardens facing the sea)
- Parks and gardens (grassed areas with formal flower beds open to the public)
- The surrounding countryside (the South Downs which act as a back-drop to the town).

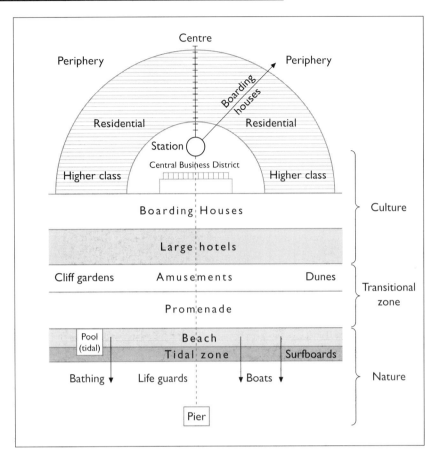

FIGURE B2.1 The North European coastal resort (after Jeans)

As a result of this, planned redevelopment of the Lanes area was abandoned and the council decided to promote plans to build a marina to the east of the Palace Pier (see Figure B2.3).

This roused strong views both in favour and against the project. Sir John Betjeman wrote: 'I fail to understand why those who defend this piece of developers' greed regard such expansion of Brighton as inevitable. No one who has the misfortune to live in Brighton will think that a garish pleasure slum built on the water will benefit the town.' By 1976 the breakwaters enclosing 77 acres of water had been built. The development also comprises flats, houses, shops, cinemas and car parks, in addition to berths for over 1000 boats. In that the Marina caters specially for affluent visitors, it gives the town an added dimension to its varied amenities, but the vast majority of visitors who come still visit the traditional tourist centres in the Lanes, the Pavillion, the Pier and the beach.

The continuing debate about the redevelopment of the Churchill Square shopping centre shows how complex the process of re-invention is. Churchill Square, a pedestrian precinct incorporating chain stores, specialist shops, supermarkets and car parks, was 'carved out' of the streets and terraces to the south west of the railway station in the 1970s to provide a town-centre shopping area. By the mid-1990s it was shabby, exposed, vandalised and its lower shopping levels had been abandoned to tramps and the homeless.

Town centre shopping had been superseded by the out-of-town hypermarket. The redevelopment package finally accepted by the town council and funded by the site owners, an insurance company, has modernised and enclosed the site and increased the number of restaurants, take-away and pub outlets. Brighton hopes it will still be able to retain its reputation as a regional shopping centre alongside its traditional attractions for visiting tourists.

NUMBER	1938	1998	NUMBER	1938	1998	NUMBER	1938	1998
1–3	Queens Hotel	Queens Hotel	54	Restaurant	Accomm.	93	Residence	
4	Café	Nature Company	55	Antiques/Res.	Accomm.	94–99	The Grand Hotel	Hotel
5		Beauty Clinic	56	Jewellers	Confectioner	CANNON PLACE		
6		Restaurant	57	Fancy goods	Take Away	102	Residence	Restaurant
7	Antique Dealer	EMPTY	58	Ice cream parlour	Camera shop	106	Chemist	Night club
8		Bookshop	59	Hotel	Night club	107	Garage	Hotel
9		Oriental Rugs	61	Hotel		109	Furniture shop	
10	Restaurant	Restaurant	62	Antiques		110–121	Hotel	Metropole
11	Motor Insurer		64	Hotel	Belgrave	QUEENSBURY MEWS		
12	Blouse shop		65	Hosiers		122	Restaurant + 4 res.	Restaurant
13	Café		66	Confectioner	Hotel	123	Wine shop/dentist	Hotel
15	Hotel	Hotel	WEST STREET			124/125	Residence	Hotel
16	Hotel/Antiques		68	Camera shop	ODEON	126	Hotel	Hotel
18	Boarding House		69	Tobacconist		127	Residence	Restaurant
19	Newsagent		70	Confectioner	CINEMA	128	Hotel	Restaurant
LITTLE EAST STREET			71	Furriers		129/130	Hotel	Residences
20/21	Watchmaker	Brighton	73	Antiques		REGENCY SQUARE		
21	Res/Psychologist		74	Ice cream parlour		131	Jewellers	Residences
22	Costumier	Thistle	76	Jeweller		132	Restaurant	Residences
23	Confectioner	Hotel	77	Dressmaker	BRIGHTON	133	Residences	Restaurant
MARKET STREET			80	Furrier		134	Residences	Residences
24-25	Hotel	Brighton	81, 82	Restaurants	CENTRE	MONTPELIER RD		
26	Stationer		82	Restaurant		143/145	Hotel	Hotel
27	Tobacconist	Thistle	84	Ice cream parlour		146	Residences	Residences
28	Hotel		85	Theatre		147	Restaurant	Residences
29	Hosier	Hotel	86	Chemist		148	Hotel	Hotel
30	Wines/Spirits		RUSSELL STREET			NORFOLK STREET		
BLACK LION STREET			87	Boarding House	The	149/150	Hotel	Residences
31	Motor Engineer	Old	88	Costumier		151/152	Boarding house	Residences
32	Confectioner	Ship	89	Tourist agency		153	Residences	Residences
33–38	Old Ship Hotel	Hotel	90	Confectioner	Grand	WESTERN STREET		
MIDDLE STREET			91	Restaurant				
51–53	Costumier	Student accomm.	92	Nursing home				

FIGURE B2.2 Changing amenities in Brighton

FIGURE B2.3 Brighton Marina

Spain

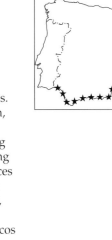

In the last 50 years the type of person visiting the Mediterranean coast has changed from one extreme to the other. Little was known of this region until the 1950s. Most travellers going to Spain were attracted to the culture and architecture of Madrid, Seville and Cadiz. The hardy explorers who reached the tiny fishing villages and small town holiday resorts on the Costa Blanca and the Costa del Sol would have found the Mediterranean staples of sun, sea, sand, friendly and hospitable local people, cheap food and wine, some Spaniards having a family holiday ... and little else. This was a classic location for an allocentric holiday!

Then the advent of jet aircraft in the 1960s bringing hundreds of tourists at a time to coastal airports led to the construction of massive high-rise hotels, financed by the tour operators, in the middle of the fishing villages and tiny resorts. They offered standard accommodation and toilet facilities, that is home-from-home comfort, with a dash of the unfamiliar – a flight to a foreign country, a different climate, different food. These were ideal conditions for midcentric tourists! Surveys showed that a typical visitor spent 26 per cent of the time on the beach, 30 per cent in and around the accommodation, 22 per cent of the time shopping and 14 per cent on entertainment. The town centres grew rapidly as a host of enterprises opened to meet the needs of the visitors.

By 1980 there were over 10 million visitors annually to the Spanish Costas. The majority came from the cold climates of industrial northern Europe, especially Germany and Britain. Hundreds of thousands of them decided to stay permanently! Property developers purchased any available land in the vicinity of the coast. Planning regulations were lax or non-existent. Mile upon mile of villas, bungalows and small apartment blocks, each with its obligatory swimming pool set in marble tiling, spread like a pink and white rash along the dusty coastal zone from Barcelona in the north, to Algeciras in the south. Many of these properties were bought by Spaniards whose income was steadily rising as a result of the tourist boom, but the majority belonged to citizens from northern Europe. They are used for extended holidays, for letting income and ultimately as retirement homes. Spain had finally become a Psychocentric location, not just home-from-home but home itself!

By 1996 there were over 50 million tourists coming to Spain annually. They stay mainly in self-catering accommodation (there are over 9 million bed spaces in private accommodation and just over 1 million hotel beds). They drink in bars run by expatriates, eat their national dishes in one of the 50 000 restaurants, spend their evenings in the clubs, discos and casinos of the resorts and their daytime hours at the aquatic parks, swimming pools and beaches. Places such as Benidorm, Torremelinos and Lloret del Mar are completely dominated by tourism as the map of Torremelinos (Figure S2.1) shows.

Local people have mainly moved away from the coastal zone to the relative tranquillity to be found a few kilometres inland. They could not escape the problems occurring as a result of inadequate planning such as the insufficient water supply and inadequate sewage facilities. Traffic was a nightmare, particularly along the notorious coast road. Long-standing industries such as fishing and agriculture were in decline as their basic resources – land and water – were allocated to tourism. The tourists themselves spent relatively little, on average US$512 a head in comparison with the world average of US$581. The major tour operators continued to extract their profits from package tours by ever greater economies of scale. The decline in the number of tourists visiting Spain in the late 1980s and early 90s reflected the fact that saturation

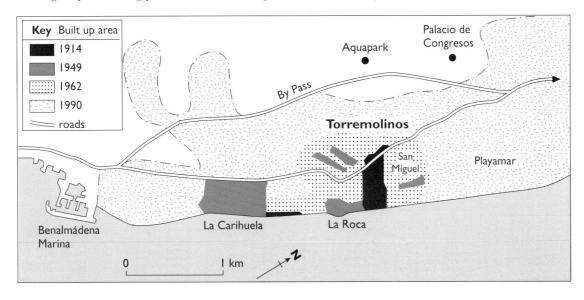

FIGURE S2.1 The growth of Torremolinos due to tourism

point had been reached. Much of the infrastructure was old and shabby and the tourists themselves had become more allocentric; they were seasoned travellers and wanted to go to more exotic locations. During the 1990s the Spanish government focused its attention on how to 'reinvent' its tourist industry. The Spanish Minister for Trade, Industry and Tourism was instructed to 're-position' Spain as a destination. They would offer cultural and city tourism, conference and conventional tourism, high-income leisure pursuits such as golf, polo, big game hunting and fishing at new resorts like La Manga del Mar Menor (see Figure S2.2).

The vast majority of visitors will still be drawn to Spain for the traditional low cost pleasures of sun, sea and sand. The Costas remain the number one leisure region for industrialised Europe.

FIGURE S2.2 La Manga del Mar

The Gambia

In the less developed world the development of a tourist industry has seemed like an appropriate way of earning much needed foreign currency. The effect that the introduction of foreigners, hotels, transport infrastructure and leisure facilities has on the country depends on decisions made in the first instance as to where the new industry should be located. If the government decides, as in Senegal and Indonesia, to allow random location, then both hosts and visitors will see at first hand how the other group lives. Alternatively, as in Tunisia, the hotels and other facilities can be constructed away from the local population, so that tourist ghettoes are the result.

In The Gambia the main holiday hotels are located along the Atlantic coast to the west of Banjul. This was the most suitable place to appeal to Northern Europeans searching for sun, sea and sand and was not a deliberate choice to segregate tourists from residents. However, there are one or two hotels close to the capital and a few up country. There is one at Mansa Konka which is a safari camp catering for Western tourists (see Figure G2.1).

FIGURE G2.1 Hotel location in The Gambia

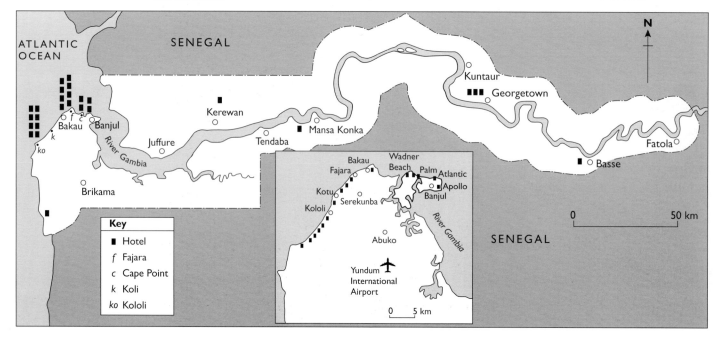

Scandinavian tourists were the first to arrive in the mid 1960s, seeking relief from the long cold winters of northern Europe in the warm, dry conditions of The Gambia. These allocentric, explorer types were happy to meet and mix with the local people. The Gambia rapidly developed a reputation for sex holidays for middle-aged ladies! Since these beginnings the more conventional tourism has expanded. Large hotels were built to cater for the growth in demand. In 1967 there were only two hotels with 52 beds in the whole of the country; by 1997 this had increased to 24 hotels with 6000 beds in the vicinity of Bakau (see Figure 2.2b). These have been designed to provide package tourists with the accommodation they would expect and to blend in with the scenery by imitating colonial or African architectural style. Although close at hand, they are a long way from the corrugated squalor that makes up most of the housing for the residents of Banjul or the simplicity of the rural villages (see Figure G2.2a).

Inevitably this disparity must cause some envy and discontent on the part of the local people and has been described as a form of neo-colonialism.

FIGURE G2.2A Rural homes in The Gambia

FIGURE G2.2B Typical hotel styles in The Gambia

Elsewhere throughout the country there are still only 350 bed spaces in a scattering of hotels and safari camps.

The airport at Yundum, 2 km south of Banjul, was originally laid out by British army personnel to provide cover for military shipping convoys travelling around the Cape of Good Hope during the Second World War. In the 1960s, during a time of strict control on air passenger services, British Caledonian, a rival company to BOAC (a forerunner to British Airways), received permission to operate a weekly air service between London and The Gambia. They built the first two hotels at Bakau in order to fill empty seats on their flights, and so the holiday industry was born. The military runway was upgraded with a concrete runway which was extended to full internatonal length in 1989 when The Gambia was chosen by NASA (North American Space Administration) as a recovery base for rockets and satellites that had ditched in the Atlantic. The essential airfield infrastructure was in place although air terminal arrangements were very basic and have only recently been upgraded (see Figure G2.3).

FIGURE G2.3 Improved airport facilities

FIGURE G2.4 Children begging from tourists by the roadside

Elsewhere in the country, transport infrastructure is at the same level. Altogether there are only 450 km of tarmac road. The remaining 4000 km have a laterite (baked tropical soil) surface which is filled with potholes and becomes impassable during the wet season. Ferries do operate along and across the Gambia river, but conditions are unsuitable for any but hardy tourists. The majority will prefer to travel around on especially chartered boats and coaches if they are to see anything of the country outside the capital.

There they can enjoy other tourist activities such as bush and safari visits or an 'Africa experience', which includes a traditional meal and native dancing in a reconstructed village. Within Banjul there are street markets and craft markets from which locally made souvenirs in carved wood and dyed cloth can be bought. However, many package tourists find it easier to stay within the confines of their hotel with bar, pool and friendly service at hand, than face the hassle and incessant demands of the ordinary people who see the European tourist as a source of largesse. The children ask continually for sweets, pens and money which they have received from visitors in the past, while young men and women are avid to meet and socialise with visitors from western Europe in the hope of establishing contact and friendship.

After the publication of Alex Haley's book *Roots* in 1976, thousands of black Americans arrived in The Gambia to see the place from which their ancestors had been sold into slavery. In particular they visited the village of Juffure on the Gambia River, which Haley had identified as the birth place of his own family. Thousands of pounds have been contributed to the village economy by Haley and others. All that remains is a broken electricity generator and the shell of a mosque, while the villagers openly beg outside their huts and the children continue to pester visitors (see Figure G2.4).

One can question whether a people demeaned by visitors who look at them as if they were a zoo species is too high a price to pay for development.

New Zealand

The location of New Zealand on the south west edge of the Pacific Ocean, three hours flying time from Sydney, has been the dominant factor controlling both settlement and tourism in the country. It is far away from the main centres of population, cut off from the migration routes and from any large influx of tourists from abroad. The tourist industry catered for domestic holiday trips in the first instance. New Zealanders rapidly discovered the astonishing variety of natural landscapes within their country and spent their annual fortnight's holiday enjoying the outdoor activities associated with mountains, the coast, rivers and lakes. At this time there was little urban development and in any case, the towns that were growing into cities, Auckland, Wellington and Christchurch, did not have the historic attractions of cities elsewhere in the world. The infrastructure needed to cope with the small numbers involved was minimal. Family hotels, camp sites or hostels provided accommodation. Food was available from cafés, pubs and restaurants.

The area of thermal hot springs around Rotorua in North Island had also become the home territory of one of the more dominant groups of Maori settlers in the fourteenth century. Their total numbers have fluctuated wildly since they arrived in New Zealand. They were an aggressive people, whose main occupation both before and immediately after European settlement was inter-tribal warfare, with the loser going to the cooking pot!

With the advent of European diseases, alcohol and weaponry, the total Maori population declined rapidly and was only 42 000 at the beginning of this century. However, the population has increased over the last 30 years and there are now 280 000 Maoris living throughout New Zealand, many of them in the Rotorua area.

Tourism had started in Rotorua in the latter decades of the nineteenth century when a road was built from Auckland in 1883 and a railway link was completed in 1894. It became fashionable to go to Rotorua to bathe in the hot springs and to drink the sulphurous water in the hope of curing various diseases. The hot springs were centred on Whakarewarewa, which was also a Maori settlement so many of the visitors also encountered the distinctive Maori culture for the first time (Figure NZ2.1).

Each Maori village or township has a meeting place or *Marae* where formal ceremonies take place. The Maori people have many traditions. They have their own language, craft forms and dances which they display to visitors at concerts held daily in concert halls and hotels. In the latter, special Maori evenings combine dancing with Maori cooking, a *hangi* feast is cooked in an earth oven underground. Meat and vegetables are wrapped in wet cloths, then steamed over hot stones placed in the confined space. Now tourists from all over the world look forward to visiting a Maori village as part of their New Zealand holiday.

For the Maori people themselves it is another opportunity to emphasise their presence on the islands. With the growth in their numbers over the past three decades they have become more confident about their identity and their role in New Zealand society. The nightly concerts and performances of Maori dancing are seen as an opportunity to display their culture. It provides Maori teenagers both with a source of money and a chance for the old traditions to be passed on to the next generation and is seen as a privilege rather than a belittling experience.

FIGURE NZ2.1 The Whakanewanewa entrance to a Maori meeting place

The Disney Concept

Visitors' maps to Orlando, Florida show clearly just what impact the tourist industry has had on the town once a centre for the orange farms of central Florida (see Figure D2.1).

Brochures state that 'Orlando is located approximately in the centre of Florida. This convenient location makes it the ideal vacation spot in Florida. Besides being home to 66 exciting attractions, more than 84 000 hotel rooms and 3000 restaurants, its central location affords visitors easy access to both the Atlantic coast and the Gulf of Mexico beaches'. The international airport now handles more than 18 million passengers a year as people arrive from all over the world, drawn by the magic of a Disneyworld holiday.

What this has meant to the area in every aspect of life can be readily grasped. The town has more hotel rooms and a higher occupancy rate (89–90 per cent p.a.) than anywhere else in the USA. It is estimated that the leisure and tourism industry has generated over 50 000 jobs in the area. Disneyworld has an employee roll of 10 000 people at its Orlando site and there are 65 other attractions in the area. Today many feel that there is saturation in provision of night clubs, golf courses, water parks etc. All these employment opportunities have led to a continuous wave of in-migration for many years with a demand for housing that has caused miles of farmland to be covered in real estate. The tranquillity of Central Florida has gone for ever.

The Walt Disney Corporation saw rapid growth at the end of the twentieth century, with the opening of EuroDisney outside Paris in 1991. However, attempts to expand in America were less successful! In particular, after three cancelled projects in California, based respectively on MGM studios, a Disney Seaworld concept and a repeat of EPCOT – to be called WESTCOT – at Anaheim, the Corporation was anxious to launch a new venture in Prince William County, Virginia.

This site is 35 miles west of Washington DC, at the junction of Interstate 66 Highway and US Route 15. Plans were to develop a theme park relating to the American Civil War costing US$625 million, on 3000 acres of land outside the town of Haymarket. The project was announced in November 1993.

Disney executives had rallied support by meeting 10 000 local people between November 1993 and April 1994 in an 'outreach blitz' to the local community and business organisations. They emphasised the increased employment opportunities with 2700 jobs created in the first instance, rising to 19 000 eventually. Additionally the region would have:

- improved infrastructure and road links
- new markets and shopping malls for visitors
- new recreational and cultural activities
- annual tax revenues of US$28 million.

Their lobbying was effective and they were offered a US$163 million package of grant aid from the State of Virginia, largely to be spent on upgrading state roads, in addition to support from the National Government in Washington DC.

By mid-summer, opposition to the plan was strengthening. In particular the local residents felt the social costs were too high in terms of increased traffic (an estimated additional 70 000 vehicles a day) and increased suburban housing sprawl with associated amenities for thousands of newcomers. Their objections were summarised as 'environmental and cultural degradation';
- Increased traffic, loss of air quality
- Loss of quality of life
- Increased land values
- Low paid, seasonal employment
- Increased crime
- Degraded water resources
- Increased taxes to meet capital costs

They had no wish to live in an area like Orlando in Florida!

They were then joined by groups of protesters from National Heritage societies. The proposed site was within half an hour's drive of 18 battlefields and 64 historic sites relating to the Civil war. Although the Walt Disney Corporation offered donations, new schools, public libraries and promised a 'green' theme park, it was to no avail. The project was cancelled in September 1994.

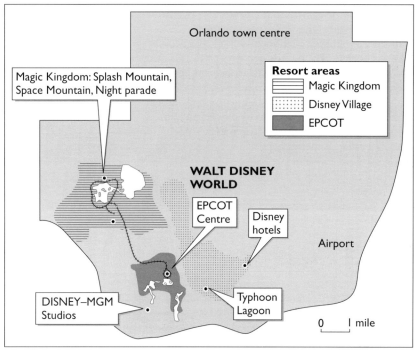

FIGURE D2.1 The impact of Disney on Orlando

3
TOURISM AND THE ECONOMY

Key Ideas

- Tourism, like any other economic activity, is subject to the laws of supply and demand
- Changes in technology have had an impact upon holiday suppliers
- Changing requirements of the tourist industry have varying impacts upon economies at both local and national levels
- Tourism plays an important part in economic growth and change both in MEDCs and LEDCs
- Styles of tourist development differ according to the source of investment
- The growth of tourism with time can be 'modelled'

Tourism as an industry

'There are many people who believe that a job in the tourism industry is in some ways less of a job than one in manufacturing'
Chairman of the English Tourist Board, Financial Times
1 July 1986

Tourism is an industry which, like many other industries, involves a large number of pieces or components being assembled together to make a complete product which is then competitively priced, advertised and sold to the consumer. However, tourism is a service rather than a tangible product. This can create a problem for potential tourists – they cannot inspect the goods before they buy but must take everything on trust while handing over substantial sums of money! This is as true for tourists returning to previously visited destinations as for those venturing into new areas because holidays are more than the sum of their individual components (accommodation, transport, attractions etc.). A faulty consumer good can be returned and replaced, but a failed holiday cannot be amended in quite the same way. Equally, the industry suppliers have the problem that their products are 'perishable', that is, today's empty aeroplane seats and vacant hotel rooms cannot be sold at a later date. Consequently, last minute bargains and heavily discounted fares are a major feature of the trade.

Despite such problems, tourism is the world's fastest growing industry and makes a considerable contribution to world economies.

Supply and Demand

Tourism, like other economic activities, is subject to the laws of supply and demand. **Demand** is the amount of a product that consumers are willing and able to purchase at a given price. Figure 3.1 shows a simple demand curve for holidays. It illustrates that demand is likely to be higher for lower priced holidays and lower for higher priced holidays.

Changes in prices are usually reflected by changes in demand although price is not the only factor which can affect demand. These other factors are:

- Consumer income: changing lifestyles in the latter part of the twentieth century have led to many people having greater amounts of disposable income. At the same time, the number of holidays taken by holidaymakers in a year has also increased;
- Changes in taste and fashion can influence where people choose to visit. For example, Turkey became a popular sun-lust destination in the 1990s at the expense of the Spanish beach resorts. (In reality, this may have been because people had become more sophisticated and confident and wished to travel further afield than the 'Costas', or because rising prices in Spain were a deterrent and Turkey therefore offered similar value for lower cost);
- Population Structure: increased proportions of people aged over 65 in Western industrialised countries have led to an increase in demand for

FIGURE 3.1 Simple demand curve for holidays

winter-sun holidays, for example in Mediterranean countries and Florida.

■ Competitors' Prices: theoretically the demand for one product depends upon the price of another. A price rise in one brand is likely to cause increased demand for other brands. However, customers may not always be aware of differences. Figure 3.2 shows the package holiday prices advertised in a brochure aimed solely at people in their 20s and those in a more family-orientated brochure. The packages appear to be almost identical – both are in the same accommodation and both offer flights from Gatwick (the 20s brochure also allows flights from other regional airports) – but there is a difference in prices. This suggests that a holiday maker would need to make extensive comparisons to find the lowest price for their desired destination.

■ Segmentation of the market became more apparent in the 1990s as the greater competition amongst the operators and increasing client awareness led to the growth in demand for specialised holidays both in terms of activities and targeted age groups.

■ Advertising can have an important effect upon demand. It is no coincidence that television audiences are 'blitzed' with summer holiday advertisements immediately after Christmas when the weather outside tends to make the thoughts of summer sunshine even more appealing. Even brief exposure on a travel programme can result in a huge rise in business, especially for small companies.

■ Legislation by governments may also affect demand. Deregulation was a deliberate government policy to reduce state control over airlines. This ended 'fare-fixing' on certain air routes and air passenger traffic grew as a consequence. Conversely, when Spain joined the EU in 1986, employers' social security taxes rose and the increase was passed on to holiday makers. The number of visitors to Spain fell as a result.

FIGURE 3.2 Same holiday, different prices

For UK airport details see page 204	Central Park Apartments		£10 UK departure tax per person included	Accommodation	CENTRAL PARK APARTMENTS		
Prices based on	1 Bedroom, 4 sharing			Code	1004		
Holiday Number	SS 4829		FREE Children at Niko	Prices based on	4 Share		
Board Basis	Self Catering			Room Type	1 Bedroom		
Departures on or between	Adult			Board	Self Catering		
	7	14	See page 4 & 5 for details	No. of Nights	7	14	
01 MAY – 09 MAY	179	189		1 May–13 May	155	165	
10 MAY – 16 MAY	185	209		14 May–20 May	175	195	
17 MAY – 23 MAY	205	229		21 May–26 May	205	239	
24 MAY – 30 MAY	265	289		27 May–5 Jun	195	225	
31 MAY – 13 JUN	225	259		6 Jun–19 Jun	215	245	
14 JUNE – 27 JUN	235	279		20 Jun–26 Jun	265	309	
28 JUN – 04 JUL	239	309	Hotel Miami	27 Jun–3 Jul	285	339	
15 JUL – 11 JUL	299	359	2nd Child pays 1st child price plus £89 per week	4 Jul–9 Jul	305	359	
12 JUL – 25 JUL	319	385		10 Jul–16 Jul	315	379	
26 JUL – 08 AUG	339	419		17 Jul–6 Aug	345	409	
09 AUG – 15 AUG	315	389		7 Aug–18 Aug	325	379	
16 AUG – 22 AUG	295	379		19 Aug–26 Aug	305	359	
23 AUG – 29 AUG	285	349		27 Aug–2 Sep	259	309	
30 AUG – 05 SEP	259	319		3 Sep–13 Sep	235	279	
06 SEP – 12 SEP	239	289	Prices are	14 Sep–22 Sep	215	249	
13 SEP – 19 SEP	229	279		23 Sep–4 Oct	175	199	
20 SEP – 26 SEP	205	259		5 Oct–15 Oct	155	179	
27 SEP – 03 OCT	189	219		16 Oct–31 Oct	135	135	
04 OCT – 17 OCT	169	209					
18 OCT – 31 OCT	165	199		Supplements per person/week	Low	Mid	High
	low	mid	high	3 Share	6.00	17.00	30.00
SUPPLEMENTS For 3	£1.00	£2.00	£3.00	2 Share	17.00	50.00	89.00
PER PERSON For 2	£3.25	£6.25	£9.25	per person in £'s based on Gatwick Sunday night departures to Ibiza	1 Share	—	—	—
PER NIGHT For 1	—	—	—		Single			
Reductions per 3rd adult £5.00 night				Supplements apply to: **Low:** May/Oct, **Mid:** June/Sept, **High:** July/Aug.			
SEASON: Low 1/5 – 23/5 & 27/9 – 31/10 MID 24/5 – 25/7 & 30/8 – 26/9 HIGH 26/7 – 29/8				(Blast off from Birmingham, Bristol, Cardiff, East Midlands, Gatwick, Glasgow, Liverpool, Luton, Manchester, Newcastle, Stansted. For full flight details see page 83.			
Remember to add flight supplements (pages 204–205) and insurance (page 207); see pages 4 & 5 for child pricing details and conditions. Booking conditions and Hint & Tips (on pages 206 & 207) which must be read before booking.				See pages 83–87 for details on insurance, flight supplements, booking conditions and holiday information. Prices are per person and may change in accordance with the 2wentys Pricing Policy.			

Supply is the amount of a product which suppliers will offer to the market at a given price. Figure 3.3 illustrates a simple supply curve.

The higher the price of a product or service, the more will be offered to the market and vice versa. As with demand there are a number of factors besides price which may affect supply. These are:
- Production costs. The CRS (Computer Reservation System) has revolutionised hotel bookings by giving global accessibility to customers;
- Production changes. Popular areas have seen an increase in the number of hotels and other accommodation;
- Legislation. Deregulation of the bus and coach industry in 1980 led to a large growth in the number of carriers and fierce competition between them and the railways;
- Firms' objectives. The desire to expand by a few suppliers in the United Kingdom led to investigation by the Monopolies Commission. This applied to tour operators, ferry companies and bus companies. It was felt their control of the market share was detrimental to fair competition.

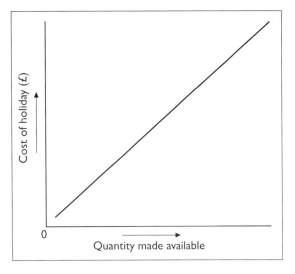

FIGURE 3.3 Simple supply curve for holidays

STUDENT ACTIVITY 3.1

1 Summarise the factors leading to fluctuations in supply and demand for tourist products.

The Suppliers

The companies involved in the provision of tourism components are the suppliers. These include travel agents, tour operators, transit operators and holiday providers.

Booking a domestic holiday in most countries is a relatively straightforward process. Tourists choose their destination, perhaps with the help of advertising and promotional literature (this is readily available for example in national newspapers and journals, from tourist information offices or in annually published guides.) A telephone conversation or letter, in some cases followed by a deposit or booking fee, is sufficient to confirm the chosen holiday accommodation. Transport to and from the destination is also relatively easy to arrange using either public or private transport as timetables and road maps can be easily accessed.

To make similar arrangements for a holiday abroad, particularly in a non-English speaking country, is too daunting for many people to contemplate. This is where the role of travel agents and tour operators becomes particularly important.

Travel Agents

These are the holiday retailers. Unlike most other retailers, travel agents do not purchase tourism products for later resale to their customers. Only when the customer has chosen their product does the travel agent make the purchase on the customer's behalf. Independent travel agents can offer impartial advice to potential customers because they have no holidays 'in stock' to try and sell.

Travel agents can arrange travel (air, rail, coach, sea and hire car), hotel accommodation and package tours. They may also offer travel insurance, currency exchange and arrange travel documentation such as visas.

The range of products that individual agents offer varies according to the nature of local demand, any specialisation of the agency (adventure holidays or business travel for example) and the preferences and marketing policies of the business owner.

Tour Operators

Tour operators are sometimes referred to as the wholesalers of the tourist industry. However, this is not entirely accurate. A wholesaler is generally someone who buys goods or services in bulk and then breaks down the purchase into smaller quantities for resale. The product is unchanged other than by quantity.

A tour operator certainly buys holiday components or elements in bulk and resells smaller quantities as required but also changes the products by assembling various elements (flight, travel to and from airport, accommodation etc.) together to form package holidays. Customers can buy the elements of their holiday separately but a tour operator is able to obtain discounts by buying in bulk.

Transit Operators

The companies responsible for transporting tourists are known as transit operators. They include airline companies, coach and car-hire firms. Air transport can be divided into three categories:
- scheduled air flights;
- non-scheduled (charter) flights;
- air taxi services.

Package holidays in particular were greatly helped by the rapid growth of charter air services in the 1960s. They are much cheaper to operate than scheduled services, which had to operate to a timetable (so any one flight may not have sold enough tickets to cover its paying passengers). They set a very high **break even load factor**. This determines the price at which seats are offered to the tour operators – being the sum of their operating costs and their profit divided by 85–90 per cent of the number of seats. Unless this number is sold, the flight owners reserve the right to cancel the flight. Charter flights don't need to be advertised to the public because the seats are sold to tour operators and in-flight services are kept to a minimum. Their biggest advantage over scheduled flights is that charter flights do not need to keep to a timetable. If one particular flight is not full enough, the passengers can be transferred to another charter flight. This may sometimes mean that potential passengers are transferred to another airport! Despite delays and inconvenience to holiday makers, passengers benefit from the lowest possible air fares.

Deregulation

The airline industry had been regulated from its early days because of:
- the need to ensure passenger safety;
- the importance of air links to national economies;
- the international nature of the industry.

This had become unwieldy and restrictive by the 1980s as a few companies controlled each route and therefore were able to charge excessively high prices by mutual agreement. The industry was deregulated in North America in 1978 and within the European Union by 1990. Prices have fallen sharply as a result; several companies have gone out of business.

Holiday Providers

This group includes hoteliers, entertainment managers and other similar roles, all of whom are providing a service to tourists. In the current segmented market, specialist companies focus on offering holidays for particular age groups.

CASE STUDY

Figure 3.4a, b, c Ria Bintan, Indonesia
a Accommodation
b Beach facilities
c Organized activities

Club Med

When Club Med first started in the early 1950s, the company offered a new type of holiday aimed at the 18–30 age group. Based originally in Mediterranean France, the complex consisted of Polynesian style accommodation in thatched roof huts, communal showers and free food, entertainment and sporting activities. Only drink had to be paid for – with different coloured beads that you bought on arrival! The simplicity proved to be very popular and the resorts – which offered round-the-clock activities – spread to other countries.

The current brochure offers over 100 locations world wide, and the range of accommodation and sophistication has increased. There are now children's facilities but the basic ideas are the same – 'A simpler world of untouched peace and beauty' (1998 brochure). Figure 3.4 shows views of Ria Bintan – the most recent Club Med resort, which is in Indonesia. According to their brochure it is, 'the glistening new jewel in Club Meds crown, with architecture inspired by fabled tropical hill resorts of old'. Ria Bintan is on an island only 30 minutes by boat from Singapore, which has one of the busiest 'gateway' airports of the world. The new complex includes 300 rooms, four restaurants, the usual sports' facilities and a range of additional activities.

The Club Med concept has recently become less popular and the company has had a series of poor annual returns. It has appointed a new management team (some from Eurodisney) and is striving to regain its touch and its profits.

CASE STUDY

Saga

Saga is a company which set up in the 1950s to fill a niche in the tour operators' market by providing holidays specifically for the elderly (those over 50). It identified the problems and needs of elderly travellers and endeavoured to meet them as part of the inclusive cost of the holiday. In particular it offers:

- Single rooms on most holidays at no extra charge. This recognises that many elderly people no longer have a partner with whom to travel.
- Budget-priced accommodation such as is available in University halls of residence during the vacations, so as to provide holidays at low cost to pensioners.
- Hosted programmes of activities in some resorts and hotels to enable single visitors to meet and socialise with each other.
- Interests and activities likely to appeal to the elderly (see Figure 3.5).
- A door-to-door service as an integral part of the holiday on more expensive trips.

Saga has identified a growing segment of the market; one with both time and disposable income for holidays. Their rapid growth is an indication of their success.

FIGURE 3.5 Some of SAGA's offered activities

Holidays with Added Interest

Holidays Exclusively for Singles
Single holidaymakers are welcome on all our holidays. All single rooms on our UK holidays and thousands in Europe are at no extra cost. Saga also offers specially hosted holidays exclusively for single travellers which include a full programme of entertainment and activities with no single supplement to pay.

Christmas Holidays
Celebrate on a Saga Christmas holiday in the UK, Europe or on a cruise.

Footsteps
Saga's Footsteps holidays with a religious theme are designed to bring the stories of the Bible and Christian traditions to life. Destinations include the Holyland, Malta and Turkey. All holidays are accompanied by a tour leader with expert knowledge and skill at bringing people together.

Special Interest Holidays
Choose from a wide variety of Special Interest Holidays in Britain and overseas, including archaeology, bowls, painting, crafts, computer courses, bridge, whist, Scrabble®, walking, golf and dancing, all led by experts.

Gardens Calendar
A collection of gardens holidays offering the chance to explore some of the most inspiring gardens in the UK and Europe. A knowledgeable gardens host will lead and guide your holiday adding valuable insights into the gardens you visit.

Music
A selection of holidays for the music enthusiast in the UK and overseas, including a Caribbean cruise with a classical music theme aboard our own ship, Saga Rose. Each holiday is accompanied by an enthusiastic Saga music host, chosen for his or her expert knowledge.

Oberammergau 2000
Register your interest in Saga's forthcoming holidays to this world-famous Passion Play.

Holidays for Groups
If you organise group holidays then Saga is able to offer you a comprehensive service. There are many benefits to booking your group holiday with Saga, for example FREE places, the opportunity to tailor the holiday to the requirements of your group and the services of a Saga representative.

STUDENT ACTIVITY 3.2

1 Use the travel section of a national newspaper and your travel agent to find the range of prices for a flight from Britain to New York. What factors influence this range of prices?

2 Study the photographs and case study information for specialised or niche holidays (Club Med and Saga). To what extent do you feel they are successful in providing for their advertised age group?

Changes with time

Tourists' demands have changed with time. Chapter 1 has considered the factors leading to the development of 'mass tourism' in the 1970s and 80s. During the mid 1970s tourism seemed to develop into an assembly line. (A parallel between the early days of mass-produced cars when Henry Ford stated that a customer could have any colour car he wanted so long as it was black, can be seen here.) Holidays became standardised and inflexible and tourists took the offered package of flight and accommodation or paid horrendously higher prices. Prices were kept low by economy of scale – the mass replication of identical holidays to an undiscerning and inexperienced clientele. The clientele were already 'conditioned' to mass production with respect to consumer goods such as cars, washing machines and televisions and so were used to owning the same products as everyone else. Holidays were just another type of consumer good to be experienced in the same way – it seemed natural to copy everyone else! It also made sense for inexperienced travellers to travel in groups with the added security of a tour operator representative at their destination. The novelty and excitement of travel were more important than the destination for people who had had no previous experience and therefore no yardstick against which to measure the quality of the service.

A generation later, tourists are more experienced, more knowledgeable and more discerning. They see the environment and culture of their destinations as an important ingredient of their holiday experience. They want a more flexible arrangement and a greater degree of independence than any mass tourism package could have offered in the 1970s. Much of this new flexibility is possible because of advances in information technology (see Communications in Chapter 2). It is allowing tourism producers to supply travel and services at competitive prices that can be tailored to individual tourism consumer needs. (In this, tourism is simply mimicking the flexible manufacturing and 'just-in-time' methods now common in the car industry. However, it has taken the car industry 60 years longer to reach this stage!) Modern cruise holidays illustrate how large ships, flexible itineraries and a range of activities on offer at each port of call can be used to produce a number of customised holidays for a large range of tourists.

Figure 3.6 shows the factors leading to the rise and then decline of old (mass) tourism in favour of new tourism.

> **STUDENT ACTIVITY 3.3**
>
> 1 In small groups discuss what is meant by each of the labels on Figure 3.6. Summarise your opinions under the heading 'Factors leading to the change in tourism over time'.

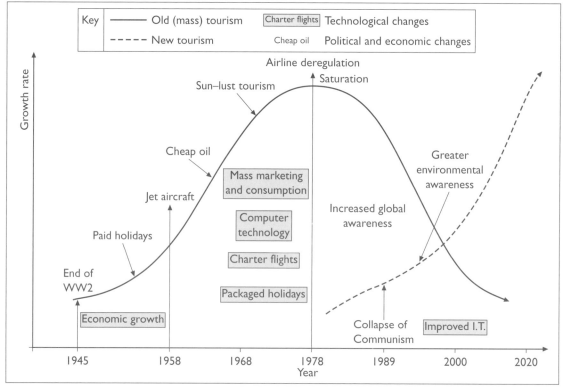

FIGURE 3.6 Changes in tourism (after Poon)

Integration

In the last 30 years of the twentieth century there has been considerable integration within and between the various sectors of the tourist industry.

Horizontal integration has taken place when two companies offering competitive products merge, for example two hotels or two travel agency chains. The merger may be voluntary or a take-over by the larger company but the result is the same. The tourist industry has become increasingly concentrated amongst a handful of major companies. The findings of a Monopolies and Mergers Commission (MMC) report were announced towards the end of 1997. It stated that travel agencies must make it clearer who owns them as consumers need to know this in order to decide whether they are being offered the fairest deal. This is particularly so where vertical integration has occurred.

Vertical integration is where companies acquire financial interests in more than one stage of an activity. Examples of this include:

- Thompson (UK) has travel agencies and transport facilities;
- Tjaereborg (Sweden) has airlines and hotels;
- Neckermann (Germany) has airlines, hotels and travel agencies.

Multi-national Corporations (MNCs)

Many multi-national companies have bought up large numbers of tourist assets world-wide. Holiday Inns have 1907 outlets throughout the world (another example of horizontal integration). They themselves are owned by Bass Charrington, a brewery group in the United Kingdom.

One of the roles of a national government is to co-ordinate economic growth so as to increase GNP. The government can encourage local industry by grants, subsidies and loans. It can attract foreign currency by favourable exchange rates and special concessions to multi-national companies. Often this appears to be the safest way of starting a tourist industry and many LEDCs have much of their tourist infrastructure – hotels, airlines, restaurants, tour operators – owned by foreign companies. In this way most of the profits earned by the venture will be passed to the shareholders. It is estimated that in 1975, 90 per cent of the hotels in the Caribbean were foreign owned. By 1985 this had dropped to 34 per cent, often by the simple expedient of 'giving' a local person paper 'ownership' of the property so as to avoid taxes levied on foreign businesses.

Governments can find themselves virtually powerless when faced by a tour operator who can divert all the potential customers elsewhere. However, with greater economic development there is strong competition from other MNCs and from local trade associations. There is also a very powerful public opinion lobby, which if aroused, as in the case of McDonalds' alleged destruction of the rainforest, can be a potent force for the control of ruthless multi-nationals.

The impact of tourism on the Gross National Product

	1985	1987	1989	1991	1993	1995	1996
				(million £)			
Visible Exports							
Electrical Machinery	8790	10 242	13 818	15 798	20 740	31 233	33 369
Chemicals	9432	10 541	12 352	13 783	17 346	21 224	22 360
Mechanical Machinery	9245	9941	12 866	13 779	15 072	18 729	20 789
Transport Equipment	6649	8636	11 065	14 025	12 682	16 379	19 989
Petroleum Products	16 114	8444	5918	6793	7801	8876	10 370
Scientific & Photographic Equipment	2987	3326	3926	4261	4770	6012	6671
Textiles	1659	1834	2205	2351	2596	3318	3472
Invisible Exports							
Tourism	5442	6260	6945	7386	9487	11 763	12 369
Financial and allied institutions	4207	5927	4183	3961	5221	6411	6397
Civil Aviation	3078	3159	3869	4039	5144	5796	6313
Sea Transport	2986	2932	3522	3351	3913	4582	4698

Source: *Office of National Statistics*

FIGURE 3.7 The importance of tourism in the British Economy

Figure 3.7 shows the importance of tourism to the British economy. In 1995 it is estimated that receipts from tourism produced £12 billion – a 200 per cent increase on the 12 year period from 1983, and 10.6 per cent of the entire 'export' earnings of the country. Many LEDCs are far more dependent on their earnings from tourism than a diversified country such as the United Kingdom. For example tourism accounts for 42 per cent of the export earnings of the Bahamas and 25 per cent of the export earnings of Jamaica. Such countries have few exports, generally consisting of unprocessed or semi-processed primary products, yet they need to purchase machinery, technology, medicines and so on from the developed world and must pay in foreign currency.

Tourism is seen as an effective route towards earning the necessary money and towards economic growth. It offers an alternative to agriculture without the complexities of setting up manufacturing industry. In particular, for small island states, tourism can quickly become the dominant source of income. The British Virgin Islands in the West Indies rely on tourism receipts for 96 per cent of their foreign income, although a nearby neighbour such as Trinidad, with a well-established agricultural base and earnings from petroleum, receives only 2 per cent of its foreign income from tourism. These figures obviously fluctuate as the prices of commodities rise and fall on the global markets. To be solely reliant on tourism is unwise; locations that are fashionable one year are neglected the year after, while events such as terrorist attacks cast a blight on the number of tourists for years after the original incident.

The impact of tourism on levels of development

Domestic tourism

As a country develops so its people will enjoy more surplus income after essentials such as food, housing and clothing have been bought. They will also have more leisure time. The seaside resorts of northern Europe saw their most rapid growth during the late nineteenth century when they became popular as a destination for a day out. Various Employment Acts which gave factory workers a week's holiday led to the provision of accommodation at these same resorts. With the advent of jet aircraft capable of transporting hundreds of passengers, people began to take foreign holidays in the mid years of the century and domestic tourism stagnated. It was only in the last decades of the twentieth century that even greater leisure time and more disposable income led to the re-discovery of home tourist locations, especially for short breaks or second and third holidays! Figure 3.8 shows how tourism grows with development.

The LEDCs are at an earlier stage on the graph. There is often little surplus income for holidays, although the resort areas of these countries will be enjoyed by local people having a day's break from work. Many migrant workers and students will travel back to their home towns and villages for any holiday period. This makes it difficult to determine how much is earned by 'Domestic Tourism'.

Domestic Tourism and the Balance of Payments

On the other hand, in many countries in the developed world there is a tradition of holiday-making within one's own country. The French do not go abroad, they go to other parts of France! Similarly, inhabitants of the other countries of southern Europe tend to take holidays along their own Mediterranean coastline. This benefits the balance of payments of the country concerned, for in some developed countries it is possible for the amount spent by people going abroad on holiday to outweigh receipts from visiting tourists.

From Figure 3.9 we can see that Spain receives a net balance of US$12 828 million, the UK has a net loss of US$1673 million while Germany has a net loss of US$15 766 million on their tourism accounts. In general, the countries of the temperate north have a deficit balance of payments, while Mediterranean and tropical countries have a surplus.

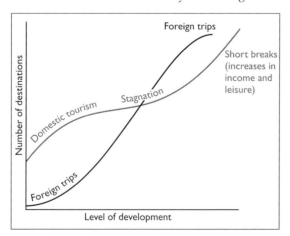

FIGURE 3.8 How tourism grows with development

FIGURE 3.9 The tourism balance of payments (OECD)

Country	Receipts (US$mill)	Expenditure (US$mill)	Balance (US$mill)
Federal Rep. of Germany	7 801	23 567	−15 766
Spain	14 780	1 952	+12 828
United Kingdom	10 196	11 869	−1 673

The Impact of Tourism on Regional Development

Many governments have used tourism as a stimulus to develop remote or peripheral areas of their country. If an out-lying region can be turned into a recreational zone for foreign visitors and for those who live in the core area, then the wealth from the core will start to 'trickle down' to the regions.

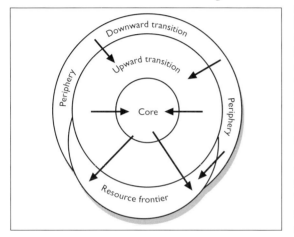

Friedmann's centre-periphery model of economic development illustrates this (see Figure 3.10).

This illustrates the cumulative effect of development suggested by Myrdal in his theory of economic growth (see Figure 3.11). Governments may encourage such tourist development, initially by provision of a transport infrastructure to open up the area then by targeted advertising to emphasise its attractions. Travel time is a crucial factor in the success of regional development plans. The Zimbabwean government wished to extend the tourist areas of their country beyond that of the attractions at Victoria Falls and Harare. They extended the runways at both Hwange and Kariba airports. In this way visitors are now able to experience a wide variety of environments within a few hours' travel of the capital city and wealthy residents can reach these remote areas for the thrills of big-game hunting (see Figure 3.12). Tourist development can sometimes provide the stimulus for much greater growth. Both California and Florida were remote, peripheral states of

Figure 3.10 (far right) Core-periphery model of economic development (after Friedman)

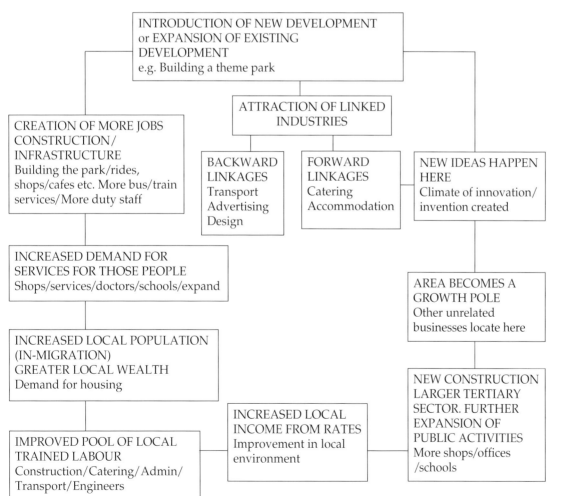

Figure 3.11 The multiplier effect (after Myrdal)

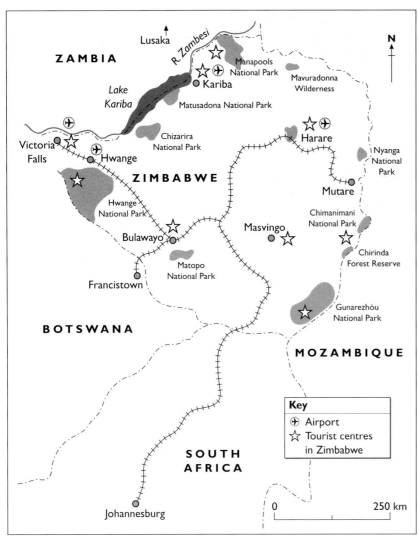

FIGURE 3.12 Regional tourism in Zimbabwe

America, far from the northern manufacturing zones. They grew rapidly as tourist centres in the mid years of the century and then attracted large-scale industrial investment for footloose industries such as electronics in California and space technology in Florida. They are now the fastest growing, most wealthy states of the country. Although they lack the resource base of other states, they can offer quality of living because of their climatic and environmental advantages.

Provision of the Infrastructure

Many governments take a financial stake in their own tourist industry by direct ownership of hotels or airlines. In this way they are able to retain more of the profits generated. They are also able to exert control over tourist development by their ownership of public utilities. Tourism needs:

- Communication links and transport facilities;
- Water supplies;
- Sewage disposal facilities;
- Power;
- Health care facilities;
- Security.

Communication and transport
Airports act as gateways to an entire region or country. An international airport, whose runway can take long-haul, wide-bodied jets, is essential to bring visitors to the country. Smaller, regional airports can enable them to reach outlying regions rapidly. Roads can then provide access to very remote areas. Tourist development in Zimbabwe typifies this.

Water
It is estimated that a visitor will require 300–400 gallons of water daily, while a golf course uses 600 000 to 1 million gallons a day. The government of Barbados decided not to go ahead with the construction of new resort complexes with golf courses when they discovered how much water would be required from supplies that were already fully extended.

Sewage disposal
Drainage networks must be efficient. Effluent should be treated before discharge. Many Mediterranean resorts lost favour with the tourist public who were not prepared to swim in inadequately treated sewage. The system of classifying beaches within the EU as Blue Cross beaches if they reach the highest standards of cleanliness and water purity has done much to focus attention on the state of the coastline (see Chapter 4). Resorts whose beaches are not up to standard are experiencing a decline in number of visitors. However, the cost to local taxes of provision of treatment plants and drainage pipes can be a major problem to small communities.

Power
Power and telephone cables must be provided.

Health care
This must be adequate for the range of activities offered to tourists in the area. For example, helicopter ambulances are needed in ski resorts, with good fracture clinics nearby.

Security
Security guards must be in evidence in particular areas where violence is widespread or where tourists seem easy prey for thieves and muggers. The state of Florida has experienced a significant fall in the number of foreign visitors following several well-publicised attacks in which tourists died. It seems that gangs of robbers were selecting hire cars as easy targets.

The above examples show that tourist development can bring problems for the local and central authorities. Roads become choked with slow-moving traffic, particularly coaches in narrow country lanes and tour buses in cities. Beaches are polluted with litter and by inadequately treated effluent. Public services are stretched to their limits during the height of the season. Public confidence is both fragile and fickle; tourists once lost, may never return.

The Impact of Tourism on Employment

The jobs directly provided in tourism are generally low paid, part time and seasonal, as Chapter 2 indicated. Highly developed countries may find it difficult to fill these positions and so use migrant labour. Many people from southern Europe have travelled north to work in hotels and restaurants. It is estimated there are 6 million migrant workers in tourism in the United Kingdom, France, Switzerland and Germany. These workers will send money home regularly; maybe they will bring their families to live with them; often they will return to their home country when they have saved sufficient to start their own business. Such migratory patterns cause social tensions both in the sending and gaining communities.

In the early stages of tourist development, costs per job are high as the employment : output ratio is high because there are few tourists. As growth occurs the employment : output ratio declines. The same number of employees can handle a much larger number of tourists. The ratio of capital investment : output is also high initially, then declines as enterprises are built and start to function. In the later stages of tourist development, when the market is saturated with shops, hotels, attractions and all prime sites are occupied, the capital : output ratio rises again. Investors have to pay large sums to purchase land or to modify land previously considered too difficult to develop on, for example marshland, or steep slopes.

The Impact of Tourism on the Local Economy

The Multiplier

At an early stage of development the **Tourist Income Multiplier** (TIM) has an impact on the local economy. Foreign exchange is used by the tourist to purchase accommodation, food, souvenirs and entertainment. Local staff are employed and will themselves spend a proportion of their earnings in the local economy. The original tourist spending has multiplied. MEDCs estimate the factor is 1.5:3.0 times the original outlay, while LEDCs have a TIM of 1:1.5.

Leakage

It is the LEDCs who experience the loss of foreign currency known as leakage. In the first instance goods and equipment to set up hotels will have to be purchased abroad if visitors are to find the same standards as they are used to at home. Commission may have to be paid to tour operators; advertising and promotion offices must be set up in the most likely foreign countries. The original tourist income is lost. Studies to discover just how much money drains away suggest that small island states suffer most (see Figure 3.13).

>90%	Mauritius, British Virgin Isles	
>50%	Less developed West Indian islands, South Pacific Islands	
>30%	More developed West Indian islands	
>10%	Developed countries importing tourist luxuries	

FIGURE 3.13 The proportion of foreign earnings lost through leakage

As a country develops, so it is able to supply more of its visitors' requirements. Figure 3.14 shows how two West Indian islands were able to reduce the amount of leakage occurring by substituting home produced goods for imported varieties.

	Barbados imported	Home produced	St. Lucia imported	Home produced
1973	66%	34%	70%	30%
1990	33%	66%	58%	42%

FIGURE 3.14 The reduction of leakage in two West Indian islands

Local producers must be encouraged to grow different crops for the tourist market. Visitors will gradually enjoy trying the local food and will not want to have everything imported from home. There will, however, always be some element of leakage and this may increase with time as the local population want to buy certain imported goods for themselves (see Figure 3.15).

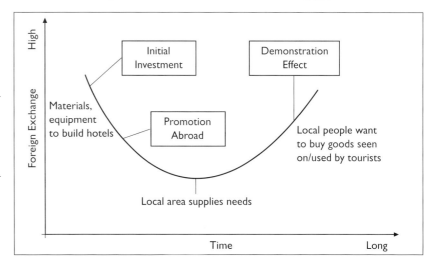

FIGURE 3.15 Diagram to show the impact of Tourist Development on foreign exchange flows of capital in LEDC

STUDENT ACTIVITY 3.4

1 a) Use the data in Figure 3.7 to calculate the % increase between 1985 and 1996 of Britain's visible and invisible exports
b) Calculate the % of total exports which tourism contributed in 1996.
c.) Write a summary statement of your findings.
2 a) Use the data given below to calculate Spearman's Rank Correlation Coefficient in order to test whether there is a significant relationship between GNP and % population taking a holiday abroad.
b) Draw a scatter graph of the data and show the line of best fit.
c) Comment upon your findings.

3 Summarise the ways in which tourism has an impact upon the infrastructure of MEDCs and LEDCs, emphasising any similarities and differences.
4 Study the cartoon (Figure Q3.4) and comment on the attitudes displayed. To what extent do you feel these views are stereotyped?

FIGURE Q3.4

Data

Country	GNP ($ per capita)	% population taking holiday abroad
Brazil	3640	1.6
China	960	0.3
France	24 990	32
Belgium	24 710	63
Ireland	14 710	65
The Netherlands	24 000	66
New Zealand	14 340	26
Portugal	9740	2
Singapore	26 730	95.8
Spain	13 580	45.1
United Kingdom	18 700	72
Zimbabwe	540	2.2

The Evolution of Tourism over Time

FIGURE 3.16 The evolution of tourism (after Butler)

Butler has attempted to produce a model showing the evolution of tourism over time (see Figure 3.16). It relates well to Plog's classification of types of tourist. Allocentric tourists are the early explorers, psychocentric types arrive when everything is consolidated! The early stages of exploration and funding are often supplied from the countries providing the tourists. The United States opened up Puerto Rico, Hawaii and Alaska. Then rapid development occurs. Tourists go to concentrated areas in the receiving countries. The tourism infrastructure – gateway airports, hotels, roads etc. – is very localised, e.g. around heritage buildings in cities, along the sea front at beach resorts, close to ski lifts in mountainous areas or along the main tourist routeways.

Over time dispersion occurs as the most accessible sites are fully developed. The more allocentric tourists move outwards to the unexplored frontiers and governments encourage spread from the core to the periphery. The core zone may go into decline or struggle to re-invent itself. Generally it will be desperate to retain its share of the market as there are huge financial sums locked into the infrastructure, which cannot be allowed to waste away.

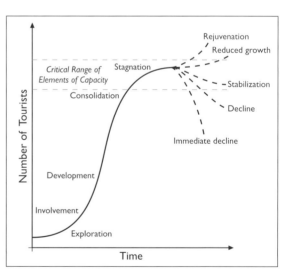

STUDENT ACTIVITY 3.16

1 Suggest a tourist area at each stage of the Butler model of the tourism life-cycle (Figure 3.16). What are its characteristics which have led to your selection?

TOURISM AND THE ECONOMY

EXAMINATION QUESTIONS

1 Assess the degree to which the tourism developed in LEDCs could be regarded as exploitative. You should illustrate your answer with a range of case studies

(15)
ULEAC

2 With reference to one or more countries in the developed world, outline how the growth of recreation and tourism has positive effects upon the economy of the country.

(9)
Cambridge

Brighton

As mentioned in Chapter 1, Brighton has hosted conferences since the nineteenth century, initially in the former stables built for the Prince Regent. A purpose-built conference venue – the Brighton Centre – was opened in 1977. It received a great deal of criticism in the local press at that time both for its cost and its appearance. The Centre cost close to £10 million, which was claimed to be a heavy burden on the town (and especially the rate payers) as Brighton Council had to pay debt charges for many years on the money they borrowed. Built on the sea front and flanked by tall, elegant Georgian buildings, the low but wide brick and glass structure (see Figure B3.1) did not find favour with the public.

Prince Charles once described the proposed extension to the National Gallery in London as a 'monstrous carbuncle', but Brighton's critics went one step further by saying that the Centre was a 'building that makes monstrous carbuncles look good'.

A further £3 million was spent on extending the Centre with the East Wing which was opened in 1991. Conferences within Brighton and Hove produce an estimated income of £75 million per year of which the Centre's contribution is £30 million.

Initially, the Centre was available for commercial lettings (that is conferences) for six months of each year and for entertainments such as concerts, exhibitions and shows for the remainder. More recently, however, Brighton and Hove Unitary Council proposed a minimum of 160 days of conferences each year. Of these between 25 and 30 have at least 500 delegates.

According to the Centre's conference guide, 'No other UK conference destination offers so much in such a compact area. There are over 2500 hotel bedrooms within a ten-minute walk of the two main conference venues. And most of the city's conference venues, hotel accommodation, shopping, restaurants, transport termini and tourist attractions are within easy walking distance of each other'

Business tourism is therefore an important component in the economy of the town.

FIGURE B3.1 The Brighton Conference Centre

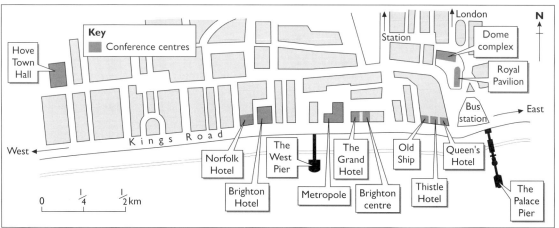

FIGURE B3.2 Conference venues in Brighton

Spain

FIGURE S3.1 The importance of tourism in the Spanish regions

Tourism is an essential element of Spain's economy. It is valued at 9 per cent of GDP and contributed 30 per cent of Spain's export earnings in the mid 1990s. This actually represents a decline from the mid 1980s when the figure was 45 per cent; a fall brought about by the growth of other industrial sectors as well as the relative failure of tourism to keep pace with economic expansion elsewhere. Tourism adds US$15 billion to Spain's balance of payments accounts with other countries each year. This amount has also declined in the last 10 years as the improvements in standards of living in Spain have resulted in more foreign travel by Spaniards and so a loss of foreign currency abroad.

Region	Number of visitors	% of tourist total	% Regional GDP
Andalucia	5.0 million	15	10
Catalonia	4.6 million	11.5	3
Balearic Islands	3.3 million	10	51
Valencia	2.5 million	7.9	9
Canary Is.	2.0 million	6	19
Madrid	3.3 million	10	3

The factors which led to Spain's rising prosperity in the 1980s were a change of government in 1986 and the decision to join the EU (European Union). Both created the conditions for the spurt of economic growth that lasted until 1991.

Although the tourist industry was forced to find increased wages and social security payments for its employees as a result of EU legislation, some of the additional money earned was spent on the purchase of holidays in Spain. An increase in the numbers taking domestic holidays had been suggested as a way ahead for the struggling tourist industry which had found itself in a state of stagnation in the late 1980s.

Not all of Spain benefits equally from the tourist growth of the last half-century. The numbers of tourists visiting each region illustrates where the inflow of tourist money has gone (see Figure S3.1).

Those areas where tourism is of greatest importance to the regional economy, such as the Balearic and Canary Islands, again show how dominant it can be in the economy of island regions or states. The figures for GDP in Figure S3.1 show this very clearly. They do not include the contribution made by tourism to air transport, building work and maintenance in these regions. Nor is the effect of the multiplier considered; that is, all the money spent locally on tourism and by its workers.

Tourism provides the principal source of employment in these areas. Vacancies for hotel and restaurant staff are filled by local people or by in-migrants from the inland areas next to the coastal provinces. Neighbouring regions then show a demographic decline, as young men and women go to the coast to find employment (see Figure S3.2).

Although jobs are seasonal and low paid, they nevertheless offer bright lights, glamour and excitement to young people from rural, inland areas and are preferred by many to the alternative, similar employment in the hotels and restaurants of northern Europe. This or rural poverty and under-employment on family farms and small-holdings was the usual choice. However, these job opportunities and migratory patterns have had an adverse effect on other industrial activities throughout the whole of Spain. Traditional industries such as agriculture, fisheries and salt-mining which provided the main forms of employment in the past are unable to attract workers today. They also find their commercial activities restricted by the demands of tourism.

Land has been sold for tourist development by landowners – often individuals or companies based in Madrid – with little concern for local impact. Property developers throughout Europe have bought and sold land in the islands and along the Costas as a speculation, in the hope of large profits. Foreign investors, in particular from Germany (who invest in the Canary Islands) and those from Britain, France and Belgium (investing on the Costa del Sol) have poured money into various schemes with the aim of making money from the sun-seeking desires of their compatriots.

FIGURE S3.2 Internal migration of population in Spain (after Pearce)

Tour operators wield immense power. They sell 10 million holidays a year in Britain and even more in Germany. Together these two countries occupy more than half the entire bed space in Spanish hotels throughout the year. They negotiate very advantageous prices from the hoteliers in the resorts because they are able to withdraw their business if they do not receive the economies of scale they seek. Then, in their home markets in Britain and Germany, the operators are able to promote their holidays by extensive advertising and glossy brochures offering 'summer sun packages' to be purchased through the outlets they also control. The existence of two or three fiercely competitive companies, such as Thompsons and Air Tours in Britain, ensures that prices are held down, while extensive 'discounting' or reduced price holidays are offered to last-minute shoppers. As far as the operators are concerned they cannot sell 'last week's holiday' so they must get what they can in order to break even.

None of these aggressive business tactics benefit the Spanish tourist industry. In particular, the 4000 hotels in the country, 60 per cent of which are independently owned, fear that the power of the tour operators may mean they are eventually taken over by foreign hotel chains. There have been various attempts to form alliances of regional owners as a consequence. The Grupo Sol Melia is an association of over 120 hotels on the Costa Del Sol, who negotiate prices with the tour operators together. In Catalonia the hoteliers and restaurant owners established an international marketing organisation in 1992 to offer theme packages at varying locations in different levels of accommodation (1 to 5 star). These have proved very successful, particularly with the short-break trade from inland cities and from the regional capital Barcelona. 1993 saw a 33 per cent increase in off-season bookings over previous years.

The rise and fall and rise again of the Spanish tourist industry relates closely to the model of the resort life cycle developed by Butler. When numbers of holiday makers become so excessive that the quality of a holiday is diminished, then a new impetus has to be found if the resort is to regain its place in the rankings of holiday location.

The Gambia

The Gambia is a very small country, 11 000 sq km in extent; with a population of only just over 1 million people, 44 per cent of whom are under 15 years of age. In these respects it has the characteristics of a small island state. The economy is also comparable to that of the West Indies or other tropical islands. Almost 74 per cent of the population work in agriculture, 3.8 per cent in manufacturing and 22.5 per cent in service industries. It is difficult for the government to diversify the economy in any way.

Groundnuts are grown as a cash crop throughout the country on land away from the ill-drained river flood-plain. They are taken to local processing factories where their oil is extracted and the residue compacted into 'cake'. Then both commodities are sold abroad, mainly to Belgium and The Netherlands, for further processing into vegetable oil and animal feed. This trade provides over 90 per cent of The Gambia's export earnings (fish – 4 per cent, and animal skins and hides – 6 per cent make up the remainder). The whole country is dependent on the quantity of groundnuts produced (which fluctuates according to the amount of rain received) and the world price (which also varies in conformity with the laws of supply and demand).

The major imports are shown in Figure G3.1. There is a perpetual deficit in the balance of payments. Imported goods generally cost over half as much again as exported products, although the transit trade to Senegal of 34 per cent of imports by value accounts for some of this difference. This discrepancy has to be met in a number of ways.

It is difficult for a country which is heavily dependent on world prices for primary agricultural products to move rapidly towards the sort of economic development that will generate wealth and progress. Without capital investment the country is caught in a poverty trap. The government can borrow capital from foreign banks, encourage investment by wealthy foreigners or look for grant aid from international sources. It was decided in The Gambia to use the tourist industry to attract foreign investment and bridge the gap in the balance of payments. Investors were encouraged by tax incentives. Imports of building materials, alcohol, soft drinks and food are allowed into the country without customs duty charges and there is no tax on profits made during the first five years of operation. Consequently, tour operators, hotel companies and wealthy entrepreneurs were encouraged to invest heavily in hotel building along the beach at Bakau, close to both the international airport and the capital.

There are now over 15 000 people directly employed in the tourist industry; this is 10 per cent of the country's wage earners. In addition a much

	%
Food	29.1
Machinery and Transport	23.2
Manufactured Goods	19.8
Fuels and Lubricants	5.9
Chemicals and related products	5.2

(Source: Gambia High Commission)

FIGURE G3.1 The Gambia's major imports

larger number of people derive most of their income from tourist spending; for example the cab drivers, craftsmen and women working in wood carving, leatherware, metalware, jewellery, batik (dyed cloth), textiles and clothing. Then there are those who earn casual money by offering small services to tourists, as guides, companions, hairdressers (plaiting Afro hair-styles) etc. The friendly, extrovert nature of the Gambian people is apparent in their desire to be of service, as is the acute unemployment problem of a less economically developed state.

The effect of the Tourist Income Multiplier, or TIM, is to spread the money earned from visitors, throughout the community. Local workers will buy food, clothing, accommodation, education and other essentials from their wages. In this way more jobs are created and those people, in their turn, will spend. Some money will 'leak' from the country to pay for imports directly required by tourists. The largest percentage of imports – 29.1 per cent – is for food, while machinery and manufactured goods together account for 43 per cent. Some of these products will be used directly in the hotels, for instance branded spirits, wines, beers, breakfast cereals and processed meats. Other goods like hire cars, tour buses, sports goods and beach furniture will all have to be imported.

In 1992 tourism contributed US$56 million to the economy, or 12 per cent of GDP (agriculture contributed 59 per cent). Imports for the tourist industry totalled US$13 million, but overall the balance of payments problem was reduced by the growth of tourism.

The Gambian government wishes to extend the industry further. Their options are:
- To increase the length of the tourist season. Currently this extends from November to May and avoids the humid, rainy season in summer. Beach holidays would not be successful in torrential rain and the inland roads are often impassable during this period.
- To spread the industry up-country into the provinces.

Most of the outlying districts of The Gambia have no input of wealth at all and desperately need an injection of capital.
- To build more hotels on the coastal strip to the west of Banjul, around Bakau.

This option may be the easiest to pursue in that the infrastructure already exists in this area. However, the remaining coastline is less suitable. Powerful Atlantic coastal currents and dense mangrove swamps rule out large tracts of land. Nor is the massive finance available to cope with immense site engineering prior to hotel construction.

The Gambia is still at a relatively early point in the tourism life cycle. More visitors arrive each year to enjoy the sun and sea and the contrasting human and natural environments, but the country has not reached the stage of development on Butler's model that is characterised by saturation economics.

New Zealand

Three quarters of the person-nights spent in hired beds (accommodation in camp sites and hostels included) are taken up by New Zealanders on their holidays – only 25 per cent are occupied by international tourists. Of these, in the early years of air travel between the 1950s and 1970s, half the visitors came from Australia and about 10 per cent from Great Britain. Many of the first arrivals were of the VFR category (Visiting Friends and Relatives). After spending a week or so with their New Zealand relations they would set out on a tour of the country, often by coach, lasting maybe for a fortnight. These early visitors set the pattern for foreign tourists which is still the blue print for the majority.

It was not until the development of wide-bodied jet aircraft in the 1970s led to long-haul travel being quicker, cheaper and more comfortable, that the

Figure NZ3.1 International visitors to New Zealand 1996

Country	Number of arrivals	Percentage of total
Australia	389 581	29
Other Oceania	64 604	5
Japan	146 316	11
South Korea	74 225	5.5
Taiwan	60 960	4.5
Other Asia	127 329	9
United Kingdom	118 490	9
Germany	56 690	4
Other Europe	75 277	5.5
North America	187 810	14
South Africa	11 166	1

(Source: New Zealand tourist office)

percentage of Australians decreased, while that from Japan, America and Europe has increased (see Figure NZ3.1).

Visitors from these regions make up the majority of the world's long-haul tourists having both the wealth and curiosity to take holidays a long way from home. The percentage of visitors from Britain has stayed very constant, though its identity has changed. The New Zealand Inbound Tourist Office identifies Younger and Older Spirited Travellers or Empty Nesters and DINKYS (Double Income No Kids) in the jargon of the trade as the two main categories of visitor. Their advertising and publicity is directed towards these types of people (see Figure NZ3.2).

Both these groups enjoy the variety of landscapes and activities that New Zealand has to offer, while the older visitors are also happy to be in an environment so close to their own in terms of culture, habits and standards. They stay on average for two to three weeks, and while in New Zealand, try to see and do as much as possible so generally make a tour of the country. Along with other international tourists they have created sufficient demand for bed spaces as to need accommodation in larger hotels than were traditionally found in New Zealand. These were mainly built in the mid 1980s in the main resorts, using foreign capital, by the international hotel chains. Typically they are geared to the demands of the coach companies who provide a daily service of tours leaving Auckland and Christchurch (see Figure NZ3.3).

One of the main changes to this pattern of activity that has occurred recently, is the growth of independent travel, particularly by visitors from Britain and the rest of Europe. Now up to 40 per cent of the holidays sold are for tailor-made, individual routes booked through teletext, rather than the inclusive packages sold by the large retail chains such as Thomas Cook and Lunn Poly (who nevertheless still account for 60 per cent of the market share). There are also many Australian specialist companies targeting both independent and package tour travellers. The new tourists are using rail, hire car, camper van and internal flights to put together their own package, so as to focus on the aspects of New Zealand culture and environment that interest them most.

Younger Spirited Travellers (YST)
- **Demographics**
  - ABC1, single, 18–34
- **Core Segment Size**
  - 843 000 (growth of 5% over 1996)
- **Geographic Spread**
  - High London up-weighting
  - Slight regional skew to Anglia/South England
- **Lifestyle Habits**
  - Professional and tertiary educated
  - Active and sporty (skiing, show jumping, tennis, boating, rugby union)
  - Eat out frequently, quality wine drinkers
  - Business travellers
  - Theatre/cinema goers

Older Spirited Travellers (OST)
- **Demographics**
  - AB1, married/partner, empty nesters, 55 years plus
- **Segment Size**
  - 753 000 (growth of 23% over 1996)
- **Geographic Spread**
  - Slight London up-weight
  - Spread in pockets throughout UK – Scotland, South England, South West England and Wales
- **Lifestyle Habits**
  - Purchasers of greenhouses/motor mowers/dishwashers
  - Likely to have a building society account/high interest in finance and investment
  - Enjoy hobbies such as bridge, bird watching, sailing/yachting, golf, tennis etc.
  - National Trust membership

Rapid growth in arrivals from the younger segment suggests greater resources should be directed to that area. However, this is to be balanced against the fact that the older spirited traveller segment is growing at a faster rate (13%) than the younger spirited travellers (1 596 000 in 1996 vs. 1 409 000 in 1995).

FIGURE NZ3.2 Consumer profiles

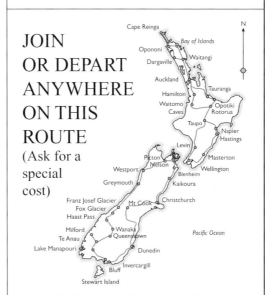

FIGURE NZ3.3 Coach company tour advertisement

They tend to stay in bed and breakfast accommodation or use motels, lodges and campsites and are less attracted to the large international hotels at the resorts. They are usually physically active and enjoy the huge range of outdoor pursuits available everywhere in New Zealand. In particular, they make for Queenstown, South Island.

This settlement started to grow in 1862 when gold was discovered, but became a ghost town by 1900 as the gold petered out. Today its situation in the Southern Alps, close to the ski-runs on the shore of Lake Wakatipu – which is fed by the fast-flowing Shotover and Karawaru rivers – provides the ideal location for access to winter sports, rafting, jet boating, bunjee jumping and flying as well as the more usual leisure activities.

Queenstown is now the tourist capital of South Island. Lots of young New Zealanders come here to work. Together with the visitors they make it a lively centre with many night-clubs and more entertainment than most of the other resorts. It is not cheap! But it is contributing a significant percentage of the 13.82 per cent of export earnings from tourism and 46 per cent of GDP.

The Disney Concept

Right from his initial idea to create a theme park, Walt Disney was conscious of the need to produce a viable business plan. He arranged a financial package with ABC (the American Broadcasting Corporation). In return for the right to show the early Disney movies such as Bambi and Snow White on television, they would put money into his venture. He was then able to buy the 75 hectares of land in Anaheim, California, on which to build his Magic Kingdom.

Unfortunately the site was so small that the company was unable to extend into accommodation and other facilities for the visiting tourists. Other people grew rich as they provided these essentials.

Subsequent ventures have been designed to allow the Corporation to build the hotels, night clubs, shopping malls and golf courses that quickly spring up in the vicinity of a Disney complex. In this way the company keeps a larger share of the takings to be made. Disney then began to plan a second theme park, to be called Disneyworld, in Florida. By now there were additional factors to complicate the business plan.

■ The Walt Disney Corporation wanted to buy 11 000 hectares of land so as to benefit financially from the vertical linkages in accommodation, shopping, transport etc. in addition to the entry ticket sales.
■ Land values would increase dramatically if it became known that the Disney Corporation was the purchaser. It was important for land to be bought at its scrubland value of US$81 per hectare. Ultimately US$5 million was paid for the site. Since then, the cost of land in Orlando has risen by 30 percent a year.
■ The total enterprise was initially costed at US$600 million. This sum could only be put together in a financial package involving other sources of money.
■ Roads in the vicinity would have to be upgraded to cope with the increased traffic flow. This involved the Government Planning Authority.

The success of Disneyworld led to the search for a site in Europe in the late 1980s. The French Government was prepared to sell 2000 hectares of land at Marne-la-Vallee, 32 km east of Paris, at the agricultural rate of US$2023 per hectare. This became one of the deciding factors in the location of Eurodisney in France rather than in Spain. Other points in the business plan were:
■ The French Government would also loan US$770 million at 7.85% interest (a very low rate for a commercial scheme);
■ Transport links were provided:
   a) to the Paris Metro
   b) to the TGV railway and the Channel Tunnel
   c) to the autoroute system and the A4/E50 Expressway;
■ Other commercial loans from banks and private companies were available to meet the initial outlay of US$ 4.2 billion;
■ The Walt Disney Corporation negotiated a complex deal with its partners whereby it received 49 per cent of the profits although its initial investment was only a fraction of that percentage.

For a variety of reasons this venture into Europe has been less profitable than the other theme parks.

The parks generally continue to be very profitable and to attract large numbers of visitors as can be seen in these 1993 figures:

Park	No. of vistors
Tokyo	15.8 million
Disneyworld Resort (Florida)	12 million
Disneyland (California)	11.4 million
Eurodisney (France)	10 million

Walt Disney identified a need and his business corporation has become expert at creating fantasy environments and thrills for millions of holiday-makers throughout the world.

4
TOURISM AND THE ENVIRONMENT

Key Ideas

- A variety of natural landscapes are exploited for tourism
- Tourist activity has a considerable impact upon ecosystems which need careful management to reduce the damage
- A variety of physical, economic, social, environmental, technological and political factors promote the development of tourist 'hotspots' in some locations
- Some regions are less likely to be developed because of a range of less favourable conditions
- The growth of the tourist industry has an important effect upon built and natural environments

Values given to different landscapes

I will lift up mine eyes unto the hills from whence cometh my help

Psalm 121

The idea that people could gain rest and relaxation from a different environment has been known for several millennia, as the quotation from Psalm 121 shows. It was not until the eighteenth century that anyone other than the explorer, trader or soldier attempted to go to these places; it was enough that they were there. Today, tourists want to visit:

- places of natural physical splendour – mountains, waterfalls, cliffs and beaches;
- places with an attractive climate – sun, warmth, snow;
- places where the plant and animal life is different;
- places where the built environment is particularly memorable.

In general, it seems that the natural environment exerts the greatest 'pull', although the creation of fantasy worlds, such as Disneyland, or artificial environments, as in the tropical savannahs of Disney Africa, show that the tourist industry can create its own world when necessary. In the past untold damage has been inflicted on some of the most beautiful places and creatures in the world, by rampant tourism. Today there is a much closer liaison between the interested parties. Figure 4.1 shows how these links have changed.

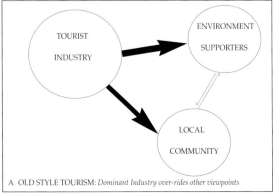
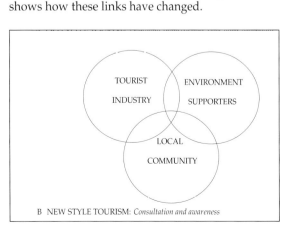

FIGURE 4.1a, b Tourism and environmental links

After Sustainable Tourism Development W.T.O. 1993

Local tourist resources

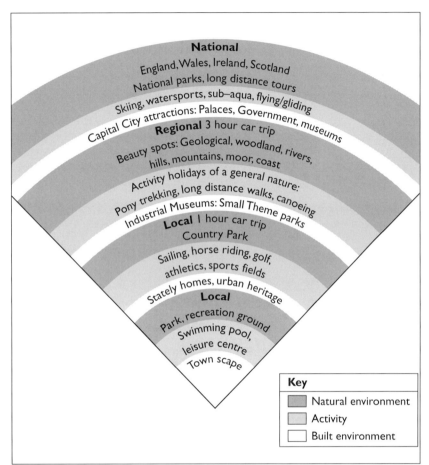

FIGURE 4.2 The range of recreational activity in Britain

Within walking distance of home, most people have a park or recreation ground in which they can gain a sense of relief from the crowded urban conditions in which many of them live. In Great Britain it is also possible to reach a wide variety of physical environments for a day trip, although all will be subject to the unpredictability of the British weather! Figure 4.2 shows the range of possible recreational activity in Britain.

- A short bus journey gives many people access to a swimming pool or leisure centre.
- An hour's car drive enables the majority of the population to reach open countryside where country parks, footpaths and common land enable the public to enjoy the varied scenery of Britain (see case study of Seven Sisters Country Park in Chapter 5). Surveys show that 21 per cent of day trips in Britain are to 'the countryside', 75 per cent of them in the family car.
- No place in Britain is further than three hours journey from the coast (see case study of Ynyslas sand dunes).
- Britain has an enormous number of stately homes and historic monuments throughout the country, which are visited by millions of people, particularly through the National Trust. This is an organisation which was founded in 1895 to protect heritage buildings and landscapes.

These local resources, whether based on the natural, or built, landscape, together make up the immense variety of locations on offer to both domestic and international tourists.

The impact of tourism on ecosystems

Each person has an impact on the places he or she visits. From the litter, graffiti and vandalism of an urban park to the rubbish strewn across the ascent routes to Mount Everest, we destroy the areas we value most. The sheer number of visitors to some beauty spots can be a form of pollution. Their transport, accommodation and waste products will further degrade the environment and destroy it for others, both now and in the future.

An ecosystem consists of a variety of living creatures and plants, **biotic elements** existing in a state of equilibrium with each other. They are supported by the **abiotic elements** or non-living elements such as soil, rain, heat and gases – nitrogen, hydrogen and oxygen. Areas which have an imbalance of relief or geological characteristics will develop their own **azonal ecosystems**. In the cool temperate climatic conditions of north west Europe, coastal areas with a surplus of deposited wind-blown sand have distinctive plants, insects, birds and mammals. They form a Psammosere or sand-dune ecosystem. Often these dune regions are popular tourist locations, particularly on the Baltic coast of north Germany and the south west coast of France, bordering the Bay of Biscay. On a smaller scale, but equally characteristic is the sand dune area know as Ynyslas, on the west coast of Wales.

STUDENT ACTIVITY 4.1

1 a) With reference to Figure 4.2, relate the range of activities in your local area to the grid. Comment on recently opened venues and those you feel are unpopular.
b) What factors cause this varying response?
c) Which has the largest environmental impact?
d) Devise a policy statement for your regional tourist board expressing what you feel should be their aims and objectives.
2 Research the range of properties and land areas owned by the National Trust. Produce a short summary of your main findings.
Does the 1988 purchase of Sir Paul McCartney's boyhood home fit in with this summary?

CASE STUDY

A sand dune ecosystem, Ynyslas, Dyfed

Figure 4.3 shows the location of the dunes. They are found on a northerly extending spit stretching across the mouth of the River Dovey (Afon Dyfi). South-west prevailing winds and dominant north-easterlies together drive the particles of gritty sand eroded from the cliffs of Dyfed, which accumulate in the form of massive dunes, starting along the beach with the small embryo dunes, then rising to fore-dunes over 10 m high. A Psammosere, or sandy environment, is created in which specially adapted plants are able to colonise and stabilise the dunes.

The embryo dunes bordering the shore support sand twitch (*Agropyron Junctiforme*), while the fore-dunes are colonised by Marram grass (*Ammophila arenaria*). These plant species are specially adapted to tolerate the sandy, windy, drought conditions. They have excessively long root systems to reach underground water. They are salt tolerant and can use the salty, brackish water beneath the dunes. Each new section of plant shoots from nodes on the stem buried in the sand. The decaying remains of old plants break down to organic matter form which then creates a skeletal soil for other salt-tolerant species. Figure 4.4 shows a cross section through Ynyslas dunes.

A succession of seral stages then develops inland across the older dunes. Annuals, perennial plants and shrubs are found, but the climax vegetation of mixed woodland is missing. The succession has been interrupted. The land has been levelled to form a large car park. Further south, constant strimming and mowing have created a short grass ley on which a caravan site is located. These amenities are provided for the thousands of visitors who come each year, from Easter to October, to enjoy a few hours or a week's holiday on a sandy beach.

Sand dunes are a particularly vulnerable physical feature. The loose, unconsolidated sand can easily be disturbed by trampling so that a blow-out occurs. The wind rips through the weak point and breaks up the dune. Sand is then carried inland and plants that have taken decades to get themselves established in the infertile soil are choked to death.

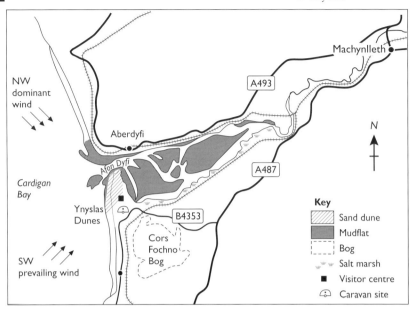

FIGURE 4.3 The location of Ynyslas

Sand dunes need to be managed effectively to minimise visitor damage.

- Facilities have been sited carefully so that routes to the shore will not follow vulnerable lines of weakness in the dunes.
- Footpaths through the dunes are protected in various ways to lessen damage caused by trampling. They are constructed of different materials depending on the stability of the sand beneath. Towards the sea where the sand is loose, the tracks usually have raised boardwalks, while closer to the car park railway sleepers have been sunk into the ground. Some heavily used areas are additionally protected with limestone blocks.
- Some dunes have been fenced off, either to give them time to stabilise or to protect nesting birds.
- Notices emphasise the potential damage to the dunes, using humour as a more effective medium than orders saying 'Don't walk here'.

It is very difficult to manage a small, highly popular location effectively so as to protect all interests. The public has a right to visit coastal areas that consist of common land. However, no one wants these places to be irreparably spoiled by the careless actions of a few thoughtless people.

STUDENT ACTIVITY 4.2

1 a) Evaluate the effectiveness of sand dune ecosystem management as outlined in the Ynyslas case study.
b) What other management methods exist to control the impact of tourism on ecosystems?
c) For an area known to you, describe how public access and impact is controlled.

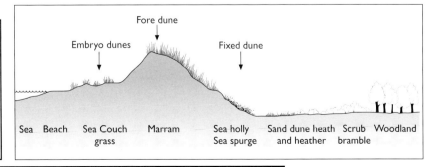

FIGURE 4.4 Cross-section of sand dunes at Ynyslas

Types of environment exploited by tourists

Each place on earth has developed in response to the particular combination of relief, rock type and climate found there. Initially we use descriptive terms like mountain, river, beach or valley, then convey a lot more information by including detail of rock type. 'Granite' mountain conveys the impression of upland plateau, while 'alluvial' flood plain suggests a wide, lowland river valley (see Figure 4.5)

It is, however, the climatic characteristics which clothe the bare bones of the landscape. The natural vegetation that develops in response to a particular range of temperature and amounts of rainfall forms a **biome**. It is possible to group places with similar climate and vegetation together in a classification of biomes of which there are nine throughout the world (see Figure 4.6)

These are all now accessible as tourist locations, though the more extreme climatic conditions such as experienced on an Alaskan trip or a trans-Sahara crossing will only be sought after by the allocentric explorer or the backpacker. The majority prefer the warm, dry climate of Mediterranean and Tropical areas, particularly in coastal resorts. Throughout the world the biggest concentrations of tourists are found in these natural environments where they are able to relax on sun-drenched beaches and swim in warm, hopefully unpolluted sea water!

FIGURE 4.5a, b, c, d, e Different landscapes

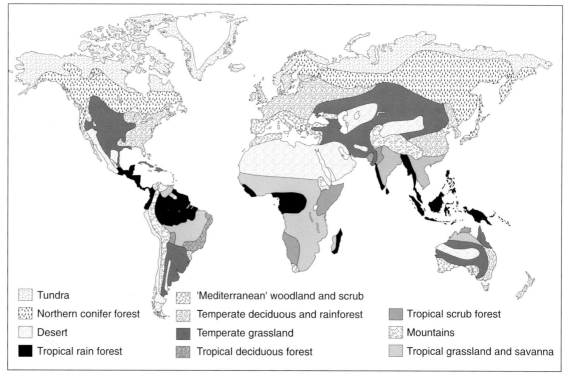

FIGURE 4.6 Major biomes of the world

- Tundra
- Northern conifer forest
- Desert
- Tropical rain forest
- 'Mediterranean' woodland and scrub
- Temperate deciduous and rainforest
- Temperate grassland
- Tropical deciduous forest
- Tropical scrub forest
- Mountains
- Tropical grassland and savanna

CASE STUDY

Tourism in the Savannah Grasslands of East Africa

A destination which is continually growing in popularity is that of the tropical savannah regions of East Africa. This biome develops in response to the high temperature and seasonal rainfall experienced on either side of the narrow equatorial belt within the Tropics (see Figure 4.7).

Zimbabwe has a rainy season that lasts from November through to April. As the sun tracks down to the Tropic of Capricorn after the Autumn Equinox, so the equatorial belt of low pressure and rain follow (see Figure 4.8). Then for the other half of the year, the country experiences a dry season of greater or lesser severity (depending on altitude and continentality) as the Inter Tropical Convergence Zone tracks northwards.

The resultant vegetation is drought-resistant woodland, shrubs and grassland. The latter grows to over 2 m tall in the rainy season, but withers and dies throughout the months of drought. Big game, lions, tigers, elephant and giraffe roam over the plains of central and east Africa alongside hundreds of thousands of deer, antelope and smaller animals. These were hunted by the pastoral tribes of the area – the Masai of Kenya and the Zulus of Southern Africa.

Since the mid nineteenth century these plains have been the destination of travellers and explorers. Initially it was David Livingstone who crossed the continent, then entrepreneurs such as Cecil Rhodes. By the late nineteenth century wealthy Europeans were arriving to experience the thrills of big game hunting. Mass tourism developed alongside the concept of the photographic safari. The clients travelled across the plains in converted Land Rovers and 'bagged' their game with the lens of a camera. Their trophies were close-up shots of lion, tiger, giraffe and elephant.

The governments of the countries involved in safari holidays have created large game reserves to protect the herds of animals from slaughter for their meat, skin or tusks by poachers. However, the policy of creating game reserves is not without problems.

- Some species overbreed, causing immense pressure on the fragile balance between species and plant life.
- Periods of drought have led to a drastic reduction in numbers of other species.
- Native resentment about land protected for game reserves has led to encroachment along boundaries.
- Poaching for ivory, leopard skins and other animal products is still happening, although there is a world ban on trade in these items.
- Licenses to operate within the game reserves as tour operators, hoteliers or guides are sought after as they are the gateway to employment and wealth.

Zimbabwe has eight main Game Reserves. Perhaps the most interesting is Hwange, which was deliberately 'created' by the government in 1929 to provide employment and a growth pole in an otherwise peripheral part of the country, where conditions were too dry for agriculture. Over 60 bore holes were drilled into the water-bearing rocks below the surface. Strategically placed salt licks also encouraged the herds of big game to come into the area. Miles of road (only 10 km of which was tarmac) were cleared and camp sites were built along the eastern edge of the reserve. This has enabled tourists to stay for a few nights in the area and provides an additional activity in the western side of the country after a visit to Victoria Falls.

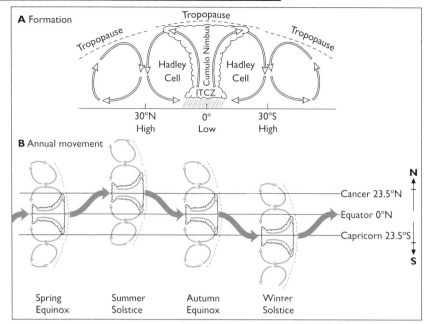

FIGURE 4.8 The effect of the inter-tropical convergence zone

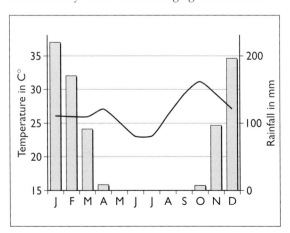

FIGURE 4.7 Climate graph for Harare

The attraction of the physical environment

STUDENT ACTIVITY 4.3

1. Study the photographs of landscape types in Figure 4.5.
a.) For each landscape state what you feel to be the underlying characteristic that has led it to become a popular tourist destination.
b.) Select one of the photographs and draw an annotated sketch to illustrate these characteristics for that area.
2. a.) Find out more about the impact that safari-style tourism has on both the economic and environmental aspects of the area involved.
b.) Discuss the pros and cons of creating artificial game reserves such as that detailed in the Disney Concept case study in this chapter.

There are also areas of outstanding relief and geology that have become centres of tourism. The extremely old rugged plateaux of Pre-Cambrian times attract fewer visitors than tertiary fold mountains. The Andes, the Alps, the Rockies, the Himalayas and the Southern Alps in New Zealand are among the most popular holiday locations in the world. These are all-year-round locations, offering a range of activities for everyone. Unfortunately the natural environment in these mountain areas has been badly degraded by the impact of thousands of tourists, as the case study of tourism in the Alps shows.

CASE STUDY

Case Study: Tourism in the Alps

The Alpine region of Europe is visited by over 100 million people each year. Sixty per cent of these tourists come during the summer months to enjoy the classic mountain scenery of snow-clad peaks, rushing waterfalls, upland meadows and tranquil lakes. As more and more tourists arrive so the natural environment suffers.

- Thousands of coniferous trees were destroyed to make way for winter sports activities.
- 40 000 ski runs have been carved out of the Alps since 1960, many in the upland areas above 900 metres where, as in Courcheval (see Chapter 2), new, high altitude resorts have been created.
- The risk of avalanches has greatly increased as land has been deliberately cleared of obstacles for ski runs, often directly above tourist accommodation.
- Water is taken from underground aquifers for use in snow machines, particularly in winter when supplies are least.
- Huge increases in energy and waste disposal facilities are required.
- Employment at the resorts offers higher pay and a more glamorous life-style than the traditional alpine occupations of farming and forestry. Consequently, these traditional industries are in decline with a damaging effect both on landscape and rural community life.
- Transport to and from the Alps is mainly by car. The meteorological conditions are often such that the exhaust pollutants are trapped in the valleys by inversion in temperature. This causes damage both to plant and human health.

Six countries share a common environment in the European Alps. Although they are good neighbours, they are also business competitors. No-one is likely to forego their market share of the tourist industry by failing to install the latest lifts, equipment and machinery. So the environmental damage continues.

In the past, the main industries responsible for environmental degradation were those associated with primary and secondary extraction and processing, such as mines and steel mills. Today tourism brings thousands of people to previous wilderness areas. Inevitably the ecological balance shifts and wild life is affected, sometimes causing permanent loss. Turtles no longer breed on the island of Zakynthos in Greece because the beaches are festooned with sun umbrellas. Climbers report that the approach routes on the major Himalayan climbs resemble the streets around a football stadium after a match – they are deep in decomposing rubbish and non-biodegradable plastics. The photograph (Figure 4.9) of holiday apartments at Puerto de Santa Maria in the Algarve is a clear example of how a landscape can be destroyed by tourism.

FIGURE 4.9 Holiday apartments on the Algarve

Hot spots and honey pots; costs and benefits

A hotspot or honey pot is one that attracts a huge number of tourists or day trippers. By definition, it is a place visited by mid-centric people, who enjoy being in a crowd. It may have a coastal or inland location and will have developed as a place of scenic beauty, historic importance or fame. The following case studies examine the environmental impact of thousands of people visiting one particular location.

Coastal hot spots

Londoners traditionally made their way to Southend, Margate or Brighton to celebrate a public holiday. Once there they indulged in paddling in the sea, sitting on the beach, going on the Pier and eating large quantities of locally caught shell fish! People of a more genteel disposition would be found at West Cliff, Broadstairs or Worthing nearby. These latter locations do not count as Hotspots. Today, the coastal hot spots are world renowned, for example Bondi beach in Sydney, St Tropez on the Côte d'Azur and the beaches of Hawaii. All have similar characteristics: beach frontage, warm water, nearby facilities, a wide range of entertainment and thousands of tourists!

Beach pollution

One of the major environmental problems associated with coastal hotspots in the last decade has been the increasing pollution experienced on beaches world wide. To combat this in Europe, the European Union awards a European Blue Flag for extremely high standards of cleanliness. Sea water is tested regularly for any trace of contamination by the bacteria found in faeces. In addition, on-land facilities such as lifeguards, warning flags, clean sand and first-aid posts are mandatory. These efforts to clean up European beaches have met with mixed success. In Britain, beaches are the responsibility of the relevant water authority, who have generally spent large sums of money to improve sea water quality, but usually not enough. Following prosecution of the North West Water Authority because of the appalling state of Blackpool beach, new sewage treatment plants were installed as part of a £150 million clear up along the north west regional coastline. However, the quality of bathing water along the region's beaches actually declined in 1998 according to the Reader's Digest Good Beach Guide (see Figure 4.10). The region currently does not have a single Blue Flag beach.

Region	Total beaches sampled 1998	Failures 1997	Failures 1998
South West	199	19	23
South East	141	10	15
East Coast	64	7	6
North West	34	14	17
Wales	171	31	24
Scotland	81	16	27
N. Ireland	26	0	2
Channel Is	24	0	0
Isle of Man	15	12	8

FIGURE 4.10 Water quality standards achieved by UK beaches

FIGURE 4.11 Weymouth, Dorset – a clean beach

Inland hot spots

CASE STUDY

Cheddar Gorge

Cheddar Gorge, a limestone valley in the Mendip Hills 65 km south west of Bristol, is an excellent example of a scenic beauty spot that has become a hot spot because so many people visit it time and again. The gorge was cut into Carboniferous Limestone by a stream which then disappeared underground. Its steep sides were shaped by the powerful downward erosion of the stream through the soluble limestone. In particular, the southern edge is almost vertical, while the northern valley side shows massive limestone blocks dipping southwards.

At the mouth of the gorge, the village of Cheddar acts as a focal point. Cafés, bars, restaurants and souvenir shops rely on the car and coach loads of visitors for their livelihood (see Figure 4.12). Most of these people will admire the gorge from the windows of their vehicle and will remain within 500 metres of the car park. Few will venture into the gorge itself on foot. Everyone is happy with this arrangement. Visitors, business people, local employees all benefit. Those people who don't like crowds stay away!

FIGURE 4.12 The entrance to Cheddar Gorge

CASE STUDY

Victoria Falls

Victoria Falls, on the border between Zambia and Zimbabwe, has become a hot spot with a global reputation. It can be classified as one of the wonders of the physical world alongside Mount Everest, Old Faithful, the Grand Canyon and Niagara. Named by David Livingstone on his exploratory journey across Africa, these immense, awe-inspiring waterfalls had been revered by natives as the location of their deity. They have formed at a point where the Zambezi River, whose catchment area covers most of Equatorial East Africa, plunges over the Great African Rift. This is an escarpment of tectonic origin running for hundreds of miles from north to south (see Figure 4.13).

European discovery in the nineteenth century led to the construction of a bridge to span the gorge below the Falls, a railway to bring visitors from all over East Africa and a palatial hotel in which to accommodate them. Today the Falls are the largest tourist hot spot in East Africa. They attract a diverse range of visitors from plane loads of middle-aged Americans to train loads of Africans and bus loads of European back-packers. The latter focus on the Falls because of the variety of activities on offer.

Companies whose young employees wear distinctive uniforms escort hundreds of people through the rapids below the Falls in 'white-water' rafts. More intrepid individuals traverse the rapids on body boards accompanied by guides in kayaks.

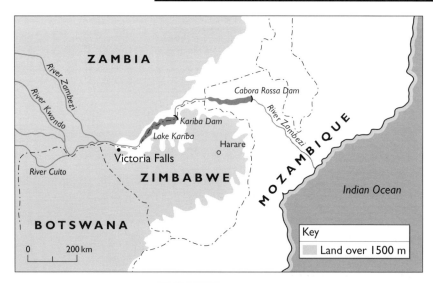

FIGURE 4.13a, b Victoria Falls

They rescue the swimmers who are unable to withstand the battering. An alternative experience is to jump from the bridge spanning the gorge on a bungee rope or to fly above the Falls in a microlight plane. The town of Victoria Falls has an air of bustling activity. There are hotels and camping sites as well as an enormous craft market, a MacDonalds, souvenir shops, travel agents, bureaux de change and so on. Many African youngsters make their way to the town, attracted by the crowds of young European back-packers. Together they create a place of excitement and adventure, of litter and lawlessness. It is far removed from the tranquility of the African Bush which surrounds it and from the magnificence of the waterfalls it depends on for existence. It is a classic honey pot.

Heritage hot spots

CASE STUDY

Stonehenge

Stonehenge, a prehistoric monument erected in Megalithic times, consists of immense stones set in a circle. The site is about 150 km west of London, on the chalk plains of Wiltshire. Its position in relation to the sun suggests that the place was one of great religious significance for the Stone Age people who lived on the chalk uplands 8000 years ago. Their skill in erecting such a temple was amazing.

In 1882, it became a designated monument under the Ancient Monuments Act and it is now the responsibility of English Heritage to protect and preserve. Large numbers of tourists visit the site. There were 640 416 in 1988, making it the third most popular historic building in England, especially for parties of school children, foreigners (who combine their visit with a trip to Salisbury Cathedral) and holidaymakers to the West Country who find Stonehenge a convenient place to stop. It lies near the A303, one of the main roads running west from London and can be seen starkly etched against the sky from the road.

This has proved to be one of the main environmental problems experienced in relation to management of the site. The large numbers of visitors need car parking, toilet and information facilities at the very least. The open nature of the site meant that they could not be placed anywhere in the vicinity of the stones without spoiling the view. Eventually the facilities were sited in a dip in the land some distance away on the other side of the road, with access to the monument via an underground walkway. The main issue revolves around the question of environmental damage set against costs of protection.

The second problem involved the long-standing tradition of the Druid religious sect of holding a special service at Stonehenge at sun-rise on 21 June. This event attracted thousands of spectators; in particular it became an annual event for travellers and hippies. By the mid 1980s there was so much aggravation, damage and disorder that access to the site at the time of the summer solstice was banned to the public except for bona-fide Druids.

Control and management of honey pots and hot spots generally rests with local authorities and business people. Neither group wish to deter visitors, because they are bound to spend money in the area, but if the quality of life for residents is badly affected, then tough controls will be introduced. People who inadvertently find they are living at a hot spot have very mixed feelings about what is happening to their lives as the following case studies show.

CASE STUDY

Gracelands

When Elvis Presley died in 1972, he was buried in the graveyard in his home town of Memphis, Tennessee, close to the mansion he had bought. Graceland, as it was known, was where he had spent his final, reclusive years. A thriving industry has been established in the area. Elvis fans can go on a tour around the mansion and grounds, visit a museum which portrays his life story, place a candle or flower on his grave and buy all kinds of memorabilia in the shops. The anniversary of his death is particularly busy when thousands of fans, from all over the world take part in a candle-lit procession and go to memorial concerts.

CASE STUDY

Althorp

When HRH Princess Diana died in a car accident in 1997, she was buried on an island in the middle of a lake situated in the family estates at Althorp, Northamptonshire – about 100 miles from London. For many years now, the Spencer family has struggled to meet the expenses associated with running an immense stately home with thousands of acres of land. Diana's brother, Earl Spencer, who inherited the title, the estate and the debts, had opened the house and grounds to visitors at £6.50 per person. Now he has converted a stable block into a museum celebrating his sister's life and tickets for entry to the house and museum (at a higher price) are usually sold out. Critics accuse him of cashing in on his sister's death, a charge he denies. 'We are doing all we can to preserve Diana's memory in fitting and respectful ways', *London Times*, 2 May 1998.

For the nearby residents in the village of Althorp, the prospect that the place may become an English 'Gracelands' with hundreds of thousands of visitors, is very daunting. The village street is already lined with 'No Parking' cones and few of the inhabitants will talk to the reporters and television crews who come to film Diana's burial place. Memories and reminiscences just fuel the interest! There is also a reluctance to be openly critical of the Earl Spencer. A resident is quoted in the *London Times*, 29 June 1998 as saying, 'We are changing from a rural backwater into a tourist trap. Like the Lord of the Manor, the earl won't let us voice our concerns'.

Conservation versus Commercialism

For many people, particularly in MEDCs, conservation of the natural environment has become an important issue. Conservation costs money and in many areas that money is generated from visitors. For example, the game reserves of Zimbabwe and Zambia are able to provide animals with a safe haven from encroaching cultivation because of the income generated by safaris. However, many of the case studies in this book suggest that tourism is a 'two-edged sword'. On the one hand it generates income and employment but on the other it can bring degradation and destruction to an environment. Generally, the degree of impact by tourism in an area is determined by the level of effective management of that area and it is therefore difficult to study the impact of tourism without also considering how any particular environment is managed. Chapter 5 examines the issue of management in tourism in some detail and includes a discussion of national parks, but the following case study illustrates how the successful conservation of one environment for visitors can have undesirable effects on another area.

FIGURE 4.14 The Great Smoky Mountains National Park

STUDENT ACTIVITY 4.4

1 a) Apply the cost/benefit formula shown in Figure 2.2 to one of the case studies outlined in the section 'Hotspots and Honey Pots'.
b) Do the costs outweigh the benefits?
c) What management problems do all these hotspots experience?
2 For another hot-spot known to you (for example in your local area), find out what elements have made it so popular and produce your own case study report for it.
3 The North Norfolk coast between Hunstanton and Cromer has many of the elements which are found at coastal hot spots. However, the area has not become highly developed for tourism. Use atlases, road maps and any other information available to offer suggestions for its lack of popularity with tourists.

TOURISM AND THE ENVIRONMENT

CASE STUDY

The Great Smoky Mountains National Park.

The Great Smoky Mountains National Park is one of 54 National Parks in the USA. It covers parts of the Appalachians in North Carolina and Tennessee and has been described as 'These mountains, mysterious and haunting are teeming with abundant wildlife, innumerable varieties of flora, and forests of spruce and hemlock' (Figure 4.14). It conforms to the United Nations 1977 definition which says that a National Park is 'a relatively large area where one or several ecosystems are not materially altered by human use and settlement; where plant and animal species, geomorphological sites and habitats are of special scientific interest, or which contains a natural landscape of great beauty; where the government of the country has taken steps to prevent or eliminate, as soon as possible, use or settlement in the whole area and to enforce the respect of ecological, geomorphological or aesthetic features, which have led to its establishment; and where visitors are allowed to enter under special conditions'.

However, as the following extract from Bill Bryson's book *The Lost Continent – Travels in Small Town America* shows, the area beyond the confines of the National Park does not appear to conserve nor respect anything mentioned by the United Nations, particularly the aesthetic!

'The Great Smoky Mountains National Park covers 500 000 acres in North Carolina and Tennessee. I didn't realise it before I went there, but it is the most popular national park in America, attracting nine million visitors a year, three times as many as in other national parks, and even early on a Sunday morning in October it was crowded. The road between Bryson City and Cherokee, at the park's edge, was a straggly collection of motels, junky-looking auto repair shops, trailer courts and barbecue shacks perched on the edge of a glittering stream in a cleft in the mountains. It must have been beautiful once, with dark mountains squeezing in from both sides but now it was just squalid. Cherokee itself was even worse. It is the biggest Indian reservation in the eastern United States and it was packed from one end to the other with souvenir stores selling tawdry Indian trinkets, all of them with big signs on their roofs and sides saying MOCCASINS! INDIAN JEWELRY! TOMAHAWKS! POLISHED GEMSTONES! CRAPPY ITEMS OF EVERY DESCRIPTION! Some of the places had a caged brown bear out front – the Cherokee mascot, I gathered – and around each of these was a knot of small boys trying to provoke the animal into a show of ferocity, encouraged from a safe distance by their fathers. At the other stores you could have your photograph taken with a genuine, hung-over, flabby-titted Cherokee Indian in war dress for $5, but not many people seemed interested in this and the model Indians sat slumped in chairs looking as listless as the bears. I don't think I had been to a place quite so ugly, and it was jammed with tourists, almost all of them ugly also – fat people in noisy clothes with cameras dangling on their bellies. Why is it, I wondered idly, as I nosed the car through the throngs, that tourists are always fat and dress like morons? Then, abruptly, before I could give the question the consideration it deserved, I was out of Cherokee and in the national park and all the garishness ceased. People don't live in national parks in America as they do in Britain. They are areas of wilderness – often enforced wilderness. The Smoky Mountains were once full of hillbillies who lived in cabins up in the remote hollows, up among the clouds, but they were moved out and now the park is sterile as far as human activities go. Instead of trying to preserve an ancient way of life, the park authorities eradicated it. So the dispossessed hillbillies moved down to valley towns at the park's edge and turned them into junksvilles selling crappy little souvenirs. It seems a very strange approach to me.'

CASE STUDY

Whale spotting in British Columbia

Following intense lobbying by environmental pressure groups such as Greenpeace and The World Wildlife Fund, international attention was drawn to the overfishing of whales and other marine mammals prior to the 1960s. Several of the whale and other species were close to extinction at the time. As a result an act was passed by the International Whaling convention banning the slaughter of whales anywhere in the world, except for a small number needed for research purposes. Japan, Norway, Canada and Russia were the chief countries to be affected by the ban. Their whaling fleets were idle and their crews were unemployed.

Today a new generation of whale fishermen has found employment as tour guides as whale numbers have rapidly increased because of their protected status. Whale spotting has become a kind of safari craze in both northern and southern waters. The table (Figure 4.15) shows the species frequenting the North Pacific coastline between Alaska and Washington state. It gives some idea how the numbers have fluctuated as various legislative acts have been passed by the Canadian and US governments. The main acts are the US Marine Mammals Protection Act of 1972 and the Canadian Fisheries Act of 1970.

Twelve-seater inflatables, based at the resorts of British Columbia and Washington state carry tourists out to sea to search for whales, dolphins and porpoises along the straits and fjords of the coast, while hundreds of thousands of people watch from the shore. This is a rapidly growing tourist activity. There were only 1400 visitors taking boat trips in 1987. By 1997 their numbers increased to 8000 around Vancouver Island, British Columbia and 80 000 throughout the whole state.

Jim Borrowman, of Telegraph Cove, NE Vancouver Island founded the Stubbs Island Charters, the first whale watching company in 1980. He says 'The whales sell. The whole area's beautiful but the main focus is the whales'.

Similar rapid growth is reported from South Island, New Zealand, while it is suggested that this could become a tourist venture for the Orkneys and Shetland Isles to the north of Scotland.

This sort of wildlife watching is infinitely preferable to the wholesale slaughter of whales and culling of seals and sea lions which took place in the name of economic necessity off the coast of British Columbia until recently. Nevertheless, there is still much unease about the excesses of some of the commercial tour operators and ferry boats as well as that of a few of the thousands of yachts, speed boats and kayaks that throng these inshore waters during the summer months. The tourists and boat people want to see the large marine mammals as close as possible, and so deliberately drive their craft into the centre of the pods (herds) of whales or zoom close to rocks, on which seals and sea lion are basking so as to watch their frantic dash back into the safety of the water. Often the cliffs are lined with hundreds of visitors who have come to spot whales from the vantage points along the coast. They report of the harassment suffered by the whales in this way. There are now suggestions that both commercial and recreational sailors and canoeists should have to attend classes in marine awareness before being issued with a license. It is felt that it would be impossible to police the hundreds of miles of coastline, so the alternative deterrent is best brought about by education!

Another controversial decision has recently been taken in the United States to permit forest fires to burn unchecked in National Parks, rather than spend huge sums of money fighting the blaze. Some forest fires start spontaneously and give rise to very effective natural regeneration of species, both of plants and animals. Visitors and tourists are horrified to encounter huge tracts of fire blackened landscape and would much prefer that the visual scenery was protected.

It often seems that actions that are acceptable to one generation are abhorrent to many members of the next. We are now amazed at the loss of landscape in the Spanish costas, concerned at the planting of conifers across great tracts of the English countryside, aghast at the reported number of animals shot by big game hunters in pursuit of a day's sport. It is more difficult to predict which of our current practices in the name of tourism will be seen eventually to be as environmentally destructive.

Figure 4.15 Marine mammals along the North Pacific coast of North America

Year	Legislation	Numbers of mammals					
		Whale			Other mammals		
		Humpback	Grey	Orca	Sealion	Seal	Dolphin
1936	Protection of Whales	23 000	200		200 000 slaughtered		
1966	International Whale Protection	1000			10 000 slaughtered		
1970	Canadian Fisheries Act			203			
1972	US Marine Mammals Act						
1997		7500	24 000	315	159 000	135 000	10 000

(Source: Beautiful British Columbia magazine April 1998)

Urban Conservation

FIGURE 4.16 Eastern Berlin

Conservation and the promotion of 'green' urban environment policies are of particular value to those who live in major conurbations. There are always two separate elements to landscape – the natural and the man-made – and in towns and cities it is not uncommon for both to take second place to economic and business priorities. The value and beauty of historic buildings mean that they should be preserved for their own worth, apart from their economic value and tourist potential. The city of Berlin in Germany is a fine example of a European city whose buildings were neglected in the eastern zone during the years of Communist rule (see Figure 4.16). The tourist potential of such a place is enormous.

The promotion of 'green' urban policies can boost both tourist and local recreational activity. Many cities are considering planting an urban 'forest' on the lines of the *stadtwalder* of German cities or the *'Bos de Amsterdam'*. Riverside walks along towpaths and through dockland areas are contributing to the clean-up of many derelict waterfronts, as in Salford, Sydney and Baltimore.

Some authorities are constructing water-parks as a leisure venue and others have urban farms. All these schemes aim to improve the quality of life for both urban dwellers and tourists.

Fortunately, it gets more difficult for any individual or company to act independently of each other. Tour operators, tourist boards and government departments all hold a watching brief. Whereas consultation often did not occur in the past, today individuals, communities and environmental pressure groups are now well able to fight for their own point of view.

STUDENT ACTIVITY 4.5

1 With reference to the case study material about the Smoky Mountains National Park and the Peak District National Park (Chapter 5), comment on which style of development you feel offers most protection for the environment.
2 To what extent should economic forces outweigh animal rights? Discuss with reference to case studies.
3 Many elements of the built environment for tourism are now showing signs of wear and tear or need the support of preservation societies if they are to survive. Suggest criteria that should be used to determine which buildings survive.

EXAMINATION QUESTIONS

1 With reference to a range of case studies, analyse the degree to which tourism can impact upon built and natural environments and show how this impact has been controlled and modified. **(25)**
ULEAC

2 Analyse the factors that cause some areas to develop as tourist 'hot spots', while other areas fail to develop as destinations for tourism. **(25)**
ULEAC

3 For one or more rural environments in the developed world, outline the conflicts which have arisen as a result of the increased use of the area for recreation and tourism. **(10)**
Cambridge

4 Describe how the expansion of tourism in the world is placing stress and pressures on the environment. **(9)**
Cambridge

Brighton

FIGURE B4.1a, b West Pier and Palace Pier

Brighton has two piers, both built in the nineteenth century for the entertainment of visitors (see Figure B4.1).

The Palace Pier is one of the top 10 non-paying attractions in Britain with its entertainments including an amusement arcade and fun-fair. The West Pier was closed in the 1970s because it was found to be structurally unsafe. A long and hard-fought campaign to save the pier from demolition, (by sea and machinery) and restore it to its former glory has been waged successfully by the Brighton West Pier Trust, but its future was not assured until March 1998, when an application for a lottery grant was finally accepted along with private funding.

The History of West Pier

West Pier was opened in 1866. It was conceived as a promenade for nineteenth century hypochondriacs as well as a landing stage for water craft and was always less raffish than the Palace Pier. It became very popular as a place to see and be seen – 600 000 visitors passed through the pier gates in 1875. By the 1920s, the original simple design of a long, uncluttered promenade leading to a theatre at the end had been altered by the addition of a concert hall, bandstand and kiosks. It remained immensely popular.

Following structural surveys made to support a takeover bid in 1965, it was discovered that the end of the pier was structurally unsafe, and the public were denied access. Rapidly increasing repair costs forced the owners into liquidation in 1975 and the pier became the property of the Crown Estates Commissions in 1978.

The Brighton West Pier Trust was formed later the same year and devoted a great deal of time and effort into raising the necessary funds to restore the pier, buying it from the Crown Estate Commission for £100. Unusually, West Pier was granted the status of a Grade 1 listed building in 1982 – the only pier in Britain to receive such an accolade. By this time repair costs were escalating – a 1984 survey estimated £7 million were needed.

In 1988, critical problems with the structure were exposed – it was in danger of collapse. An immediate £200 000 worth of rebuilding needed to be carried out or alternatively 30 metres of the pier had to be cut away at a cost of £18 000. The cause of this damage was probably the 'Great Storm' of October 1987, which had battered the south east coast of England particularly severely.

The Heritage Lottery Fund awarded £1 million to the West Pier Trust in 1996 and this enabled emergency repair work to be done. Now a private backer has invested £10 million in a scheme to restore the pier to its former role as an entertainments centre. The Lottery fund has contributed another £14 million to ensure that the pier survives.

Spain

The Restructuring of the built environment. A case study of Barcelona

One of the ways in which Spain has tried to stop the decline in number of tourists is by attracting visitors to areas other than the Costas, also by providing a range of activities targeted at different age groups and interests. One resort may now have a marina, another a golfing complex. In many of the resorts the high-rise hotels which dominated the skyline and beaches in the seventies have been replaced by three or four storey apartment blocks separated by gardens. In the cities, huge areas have been redeveloped as Spain suddenly became more prosperous after joining the EU in 1986.

Barcelona, in north east Spain, was chosen as the site for the 1992 Olympics against competition from other cities including Athens. The Committee wanted a European site for the Games after the 1988 Olympics in Seoul, and it was felt particularly appropriate to focus on Spain at that time. The country had been ruled by the right wing government of General Franco since the Civil War in the late 1930s, but with his death the king, Juan Carlos, was reinstated and an elected government took office.

Barcelona is the capital of Catalonia, a region in the north east corner of Spain, to the south of the Pyrenees. In many respects, Catalonia has a different identity from the rest of Spain, being linked to France and to northern Europe by the coastal corridor which carries the main Autoroute, the Autopista (see Figure S4.1).

Its origins go back to pre-Roman times and it has always been the centre of a thriving economic region, based initially on trade in surplus farm produce, then on textile manufacturing and heavy engineering and, in more recent years, on service industries. It is typical of many of the historic industrial cities of northern Europe in that it is characterised by central architectural splendour fringed with urban decay. In particular, in recent years, its environment has been heavily polluted by exhaust and industrial emissions and by sewage effluent. It is estimated that there is a 10 cm layer of organic mud stretching 4 km away from the coast, which is formed from human waste.

The area has a thriving tourist industry based on the coastal resorts along the Costa Brava to the north east, where there are cliffs and bays, and the Costa Dorada to the south west, which largely consists of more open, sandy beaches (Figure S4.2).

Visitors had arrived from northern Europe as part of the wave of mass tourism in the 1960s and 70s as well as Spaniards from Barcelona and the inland towns, for whom the coastal zone was their nearest seaside. However, the late 1980s and 90s saw a decline in the number of visitors to Spain from abroad, as tourists tired of tawdry, crowded conditions on the beaches and in the hotels. For officials in the Ministry of Tourism in Madrid, the successful application of Barcelona to stage the 1992 Olympics offered a wonderful opportunity to revitalise both the city and the region.

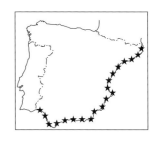

FIGURE S4.1 The position of Barcelona

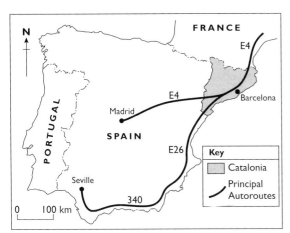

FIGURE S4.2 The coastal resorts of Catalonia

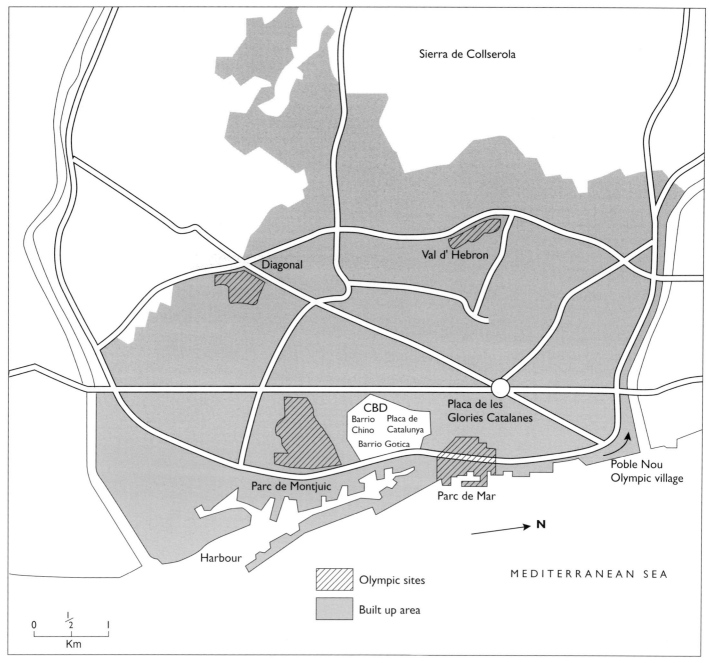

FIGURE S4.3 The main Olympic sites in Barcelona

Initially, ten locations were identified as possible sites for the various events. These were all relatively close to the city centre and were at major transport intersection nodes. Four were finally selected as Figure S4.3 shows.

The Olympic village to house the athletes was built on the coast to the north east of the city at Poble Nou. Catalonians are famed for their individualism and the housing consisted of distinctive 7-storey apartment blocks, built facing the sea in curved crescent shapes. A nearby station was renovated to provide rapid transport links with the city centre, while Barcelona airport also received a major upgrade in all its facilities. Trees, shrubs and flowers enhanced every street and square throughout the city, while specially commissioned sculptures were sited in prominent places.

Now, the Catalonians have a vastly improved regional capital with superb sporting facilities, faster, cleaner transport systems and additional publicly owned accommodation for 20 000 people. It has always had good cultural amenities with its cathedral, opera house, museum and art galleries in addition to the beach resorts nearby. In this respect the built environment has reinvented itself very successfully for the benefit of the whole region.

TOURISM AND THE ENVIRONMENT

The Gambia

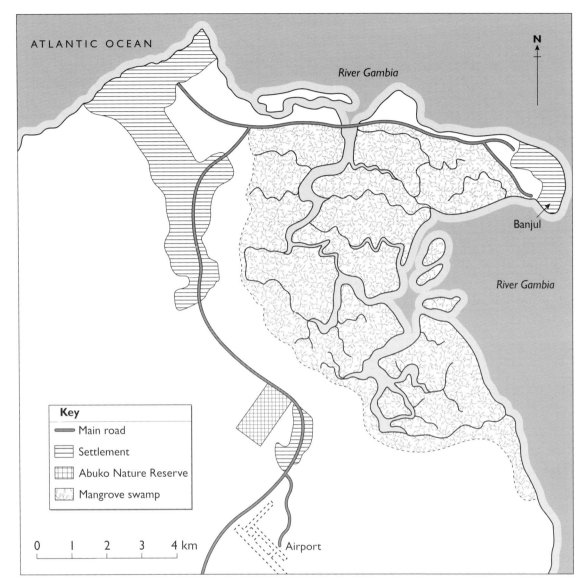

FIGURE G4.1 The location of Abuko Nature Reserve

The area south west of Banjul, close to Yundum airport, known as the Abuko Nature Reserve, was originally fenced off from surrounding land in 1980. It was set aside by Presidential decree as an area of land that represented the heritage of Gambians (see Figures G4.1 and G4.2).

It was established as a reserve in which the natural vegetation of densely wooded rainforest and savannah grassland is protected to provide a habitat for some of the indigenous species of West Africa. The whole area is very small, measuring little more than 10 km^2 in a rectangular shape (see Figure G4.3).

Nature trails have been established through the reserve along which notable species of bushes and trees have been identified and labelled, as is the custom in Botanic Gardens throughout the world. Wooden hides have been constructed overlooking waterholes so that reptiles, mammals and birds can be observed. The educational function of the reserve is emphasised by the amount of information regarding species and their habitat that is pinned to the interior walls of the hides, along with detailed safety instructions for visitors. Crocodiles, monkeys and baboons can be seen from this vantage point while a quiet walk through the reserve reveals butterflies, beetles, tree-bugs, dragon-flies as well as many other insects, lizards, snakes and a wide variety of birds.

The Gambia has always been famed for the number of migrant birds which winter in West Africa, having travelled south from Europe, in addition to domestic species. The location of Abuko between mangrove swamp, savannah grassland and rainforest, only 15 km or so from the Atlantic Ocean means that species from all three local habitats can be seen.

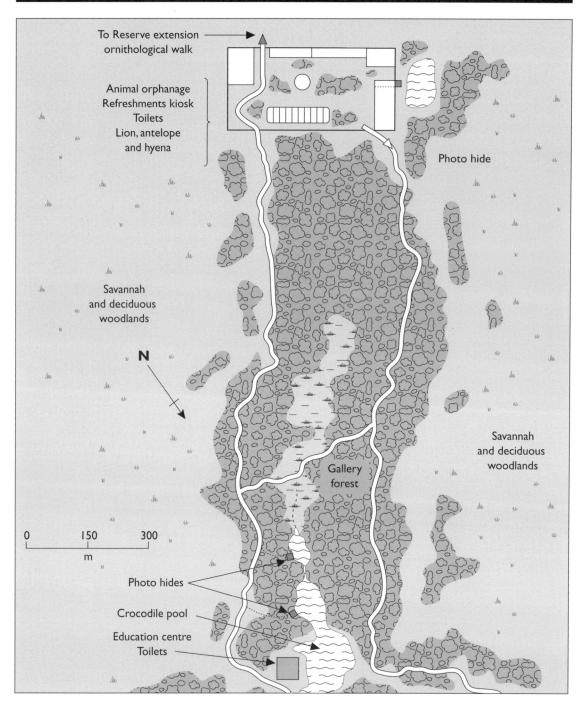

Figure G4.2 Abuko Nature Reserve
(Source: The Wildlife Conservation Department of the Gambia)

Tourists are organised to visit the reserve by their tour company representatives. They arrive by taxi or coach and spend a morning or afternoon wandering along the trails or waiting in the hides for a sight of the more elusive species. A small zoo with refreshments is provided at the far edge of the reserve, as well as a craft market at the exit. The trip provides a welcome break from the routine of beach and pool-side sunbathing and gives the visitor a taste of Africa. For the local Gambians it represents something of a problem and is a cause of conflict.

Those whose fields adjoin the reserve report that their crops are stolen or destroyed by the animals who live inside the fences but venture out to feed. Agricultural land is scarce and the reserve takes up space that could be utilised to grow food crops.

Those involved in tourism or sales to tourists in the vicinity of the reserve, would not want to see it destroyed.

It represents a small reminder of what the natural conditions in The Gambia were like before development changed the landscape, and so has value for the future. As it is so small, it cannot really claim to preserve a true rain-forest environment, but it does maintain a habitat that would otherwise be lost.

FIGURE G4.3 The Banjal Declaration

New Zealand

FIGURE NZ4.1 Centres of eco-tourism in New Zealand

New Zealand has a greater variety of natural environments and ecosystems within its relatively small geographical area than any other country in the world. From the sub-tropical rain forests of North Island to the glaciers and fjords of the extreme south, the different climates provide the abiotic conditions for a unique range of flora and fauna. A recent New Zealand tourism board slogan, 'Get as far away from it all as you possibly can', is aimed at stressed out yuppies and middle-aged executives and uses the remote location of New Zealand as its selling point. This remote location also explains the peculiar nature of the ecosystems that have developed. New Zealand is so far from the main continental land masses that living creatures created their own seral communities, including their own evolutionary characteristics. Figure NZ4.1 shows the location of Waipoua Forest Park.

This reserve has a climax vegetation of sub-tropical rainforest with ferns, vines and epiphytes. It was formed in 1952 to protect the remnant of kauri trees that had once stretched extensively throughout North Island. Thousands of visitors are encouraged to visit the forest and admire famous, named, individual trees. The kauri is similar to the Californian redwood and can grow to giant size – on average it is 30–40 metres high – and up to 1500 years old. It has been indiscriminately felled over the last 150 years for its wood, which is used in construction, boat building and furniture making. Before the reserve was created, the tree was in danger of extinction; now it has protected status and special tree nurseries propagate seedlings.

This forest is also the home of a rare New Zealand bird species, the kokaho, which, along with several others, is also rapidly declining in numbers. The variety of birds found in this country but nowhere else in the world shows the different evolutionary processes at work here. Many of the species on the endangered list are flightless. Some, such as the takahe, are almost extinct, whilst the moa, an ostrich-like bird, was completely wiped out by the Maoris.

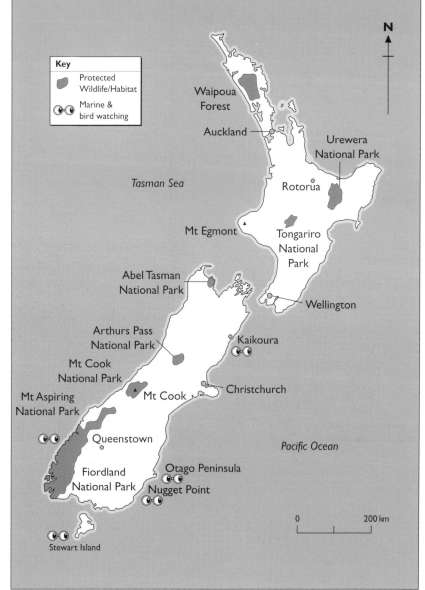

(Source: New Zealand Tourist Board)

FIGURE NZ4.2 Kauri trees in Waipona Forest

The best known of these localised birds is the kiwi, a short, tubby creature with a long bill which can be seen at 'nocturnal' view points throughout the country. Because it has become the nickname for New Zealanders, visitors enjoy going on a kiwi-sighting trip. There are however, specialised bird watching expeditions, especially to Stewart Island in the far south, where penguins, parrots and Antarctic migrants can all be seen along with the rare New Zealand species (see Figure NZ4.1).

New Zealand has few native mammals except for bats. Presumably the lack of any form of land bridge or stepping-stone islands prevented the spread of animals from Australia. It was the first Maori settlers who brought rats and a few dogs with them. Then the Europeans arrived with boat loads of farm animals, with rabbit, deer and possum on the side! The flightless birds of New Zealand had no defence against these aggressive newcomers and their numbers were quickly reduced. Much of the wild natural vegetation was also tamed as grassland was planted for the 100 million sheep, which, until recently, have formed the core of the economy centred on the Canterbury Plains of South Island as classic sheep rearing country.

The town of Kaikoura on the east coast of South Island (see Figure NZ4.1) has become the centre of marine eco-tourism in the country. Here the shore plunges to depths of over 2 km creating the ideal habitat for plankton and other fish nutrients in the upwelling waters. Dolphin, seal, sperm whale, humpback whale and minkie whale all cluster in this area because of the rich food supplies. In the past these mammals and fish were slaughtered by whalers from as far away as Japan as well as those from New Zealand ports. Today they are protected species although still at risk from over-fishing with drift nets by fleets of boats from throughout the Pacific, which gather up fish indiscriminately.

Tourists and visitors are able to watch the whales, dolphins and seals at close hand by taking a special boat trip. Wet suits, masks and snorkels are provided for those who want to swim with the dolphins. This enables people to listen to the high-pitched whistle emitted by these mammals as they communicate with each other and attempt to befriend the newcomers.

Marine conservationists are concerned that there is no management strategy to resolve the conflict between fishing and tourism or to protect the dolphin and whale from this rapidly growing activity. One researcher reported that a pod (group) of dolphins was surrounded by a constant stream of tourist craft for 8 hours 20 minutes out of a 9 hour observation period. Many of the boats are unlicensed and show little concern for the dolphins. Dr Gerry McSweeney, a New Zealand naturalist, is reported as saying 'The tourism industry should stop cashing in on the hard work and achievements of the nature conservationists and start making a substantial contribution to protect the environment' (*Independent on Sunday*, 26 April 1998).

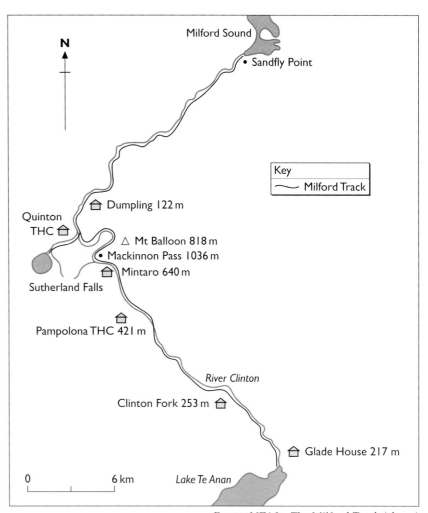

FIGURE NZ4.3a The Milford Track (above)
b Rest stop on the Milford Track (below)

New Zealand is a place where many forms of sustainable tourism have been pioneered and developed. New Zealanders are outdoor people. Their love of outdoors and simple, adventurous pleasure pursuits such as tramping has given rise to the famous tracks through the National Parks, which can be walked by enthusiasts in three or four days. Huts are provided en route for basic overnight self-catering accommodation. Such is the glamour attached to the Milford Track in South Island (Figure NZ4.3) – 'the finest walk in the world' – that it is necessary to book months in advance. Americans and Japanese tourists queue for a place on a highland walk in an area where over 5000 mm of rain falls a year. Raincoats are essential!

A tourist board slogan of the 1980s was 'Come to New Zealand, its clean and green'. This was eventually discarded because it didn't convey the excitement and exhilaration of outdoor life in New Zealand. Nevertheless, it does capture the essential quality of the country – one that can remain a selling point into the future.

The Disney Concept

FIGURE D4.1 Disney Animal Kingdom logo

'Disney's Animal Kingdom Theme Park is "A new species of Theme Park" where magic and fun meet the drama and majesty of the animal world. It's pure Disney, gone wild!'
                      Walt Disney World Resort Brochure

This £1.5 billion venture by the Disney Corporation, which opened in April 1998, is set in 200 hectares of land in the Disney World Resort in Florida. Within the area are many of the characteristic features of a modern zoo, a theme park and a safari park. These are blended together to produce the typical Disneyland experience, including the queues!

The focal point of the park is the Safari Ride. Forty hectares of land have been landscaped and planted to recreate the feel of the tropical savannahs of East Africa – the African 'bush'. Visitors board a safari vehicle in a mock African village for a 20 minute ride through roaming herds of animals; elephant, giraffe, lions, rhino, hippos and other African species. Altogether 1000 animals from 200 species have been brought from other zoos and safari parks in the USA and Europe to stock Animal World. They are kept apart by moats and other means to create an illusion of freedom while visitors endure a bumpy ride through the bush (simulated by pistons and springs in the vehicle) and then chase a gang of elephant poachers as part of their experience! The biggest problem encountered to date has been the refusal of the animals to adapt their life-style (which consists of eating at dawn and dusk, then sleeping through the day) to the needs of the tourists.

More dangerous animals such as gorillas and hippo can be viewed safely from specially constructed areas. A short train-ride on the *Wildlife Express* takes people behind the scenes to the Conservation area where they can see vets and keepers in action and get some idea of the conservation and environmental protection measures, which are being used in the park. Prehistoric time has been recreated in *Dinoland USA*, reached from the Safari Park via the Oldengate Bridge, a 12 m brachiosaurus skeleton.

Here the theme is that of an archaeological dig. Visitors are invited to search for fossil remains in a boneyard. After they have unearthed the plastic skeletons of various creatures they can get carried back in time on a white-knuckle ride – *Countdown to Extinction* – to the Palaeozoic geological era when dinosaurs were alive. The ride ends fittingly in a collision with an asteroid!

Other parts of the Animal Kingdom focus on the world of insects in *It's tough to be a bug*. This is a 3D film with intermittent sprays of hot and cold water as additional effects. It explains life from the viewpoint of an insect. A stage show entitled *The Lion King* is based on the Jungle Book and the Disney movie.

Disney 'Imageers', the engineers who created the Animal Kingdom, have successfully blurred the boundaries between reality and fantasy throughout the park, to the concern of Animal Rights groups. Although the Disney Corporation emphasises that conservation and animal specialists were consulted throughout all the planning stages, and that environmental protection is a high priority, criticism has focused on the exploitation of animals for commercial gain and the confusion created in young minds when children are presented with fiction, simulation, and flesh and blood at the same time.

5
TOURISM AND MANAGEMENT

Key Ideas

- The perceptions of tourism vary between different social groups, governments and different national and international business communities
- Different philosophies produce different attitudes and values towards tourism
- The existence globally of a wealthy 'north' and a poor 'south' and of an increasing gap between the two has led to different patterns of tourism in the developing world
- The degree of impact of tourism on built and natural environments reflects the level of effective management
- Eco-tourism and sustainable tourism development are the most likely ways forward

'... the highest purpose of tourism policy is to integrate the economic, political, cultural, intellectual and environmental benefits of tourism cohesively with people, destinations and countries in order to improve the global quality of life and provide a foundation for peace and prosperity'.

David Edgell, Chief Executive US Travel and Tourism Administration

It is the role of central government, who often devolve power to local planning authorities, to determine the land use priorities of any particular area and to reconcile conflicting demands. They must determine the crucial function of each location and what level of protection it merits. In this way a range of activities can be sanctioned, from mining and quarrying in a national park, to the provision of fun-fairs and amusement arcades on a pleasure beach. The beliefs or philosophies of a government will therefore have considerable bearing on the way tourism is allowed to develop within its country.

The impact of different philosophies

Pope John Paul II first recognised the immense power inherent in tourism as a force for good in the world in a speech to the Vatican Council in 1970. His views were expanded in the quote at the beginning of the chapter.

How and whether tourism can be used to bring about peace is largely a matter of conjecture, that it can bring prosperity is indisputable. Governments, however, may decide that the costs are too high and may restrict tourism. When communism was at its strongest in the USSR and China, and the State owned both land and all the means of production, planning and development was an element of the central economic control and that included tourism! Internal visits and trips were monitored; travel abroad was almost impossible. Holidays were an extension of the workplace, often used as a reward for meeting production targets. People who visited the USSR or China did so under the auspices of the national tourist agency of that country and were carefully shepherded along pre-determined routes. To depart from the tour itinerary was to risk police questioning and possible imprisonment.

CASE STUDY
Poland

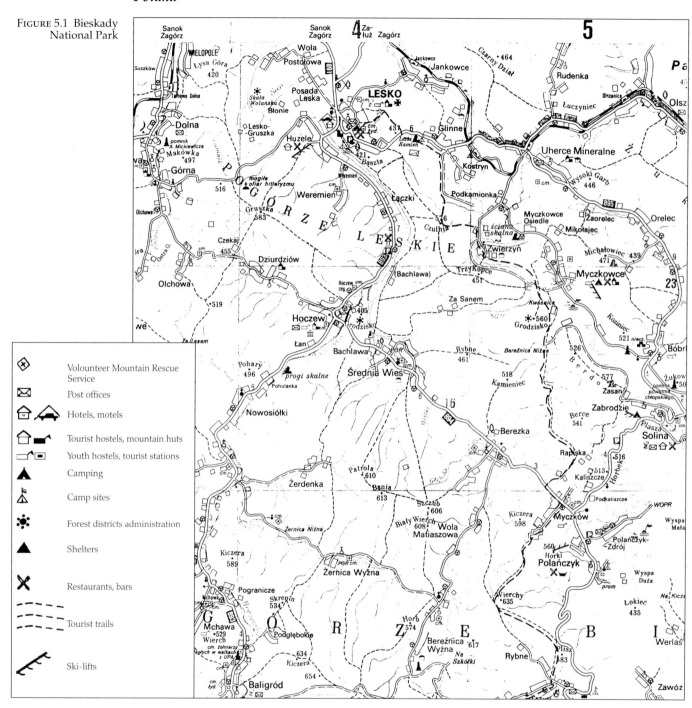

FIGURE 5.1 Bieskady National Park

Legend:
- Volunteer Mountain Rescue Service
- Post offices
- Hotels, motels
- Tourist hostels, mountain huts
- Youth hostels, tourist stations
- Camping
- Camp sites
- Forest districts administration
- Shelters
- Restaurants, bars
- Tourist trails
- Ski-lifts

With the collapse of communism in 1990 those eastern European states which had been in the economic and political control of the USSR were free to develop their own future as they wished. Prior to 1990 travel and tourism had not ranked very high on the list of economic priorities either for the Polish government, or for visitors to Poland, mainly friends and relatives. The state owned all types of accommodation and all forms of transport, including Orbis hotels and Lot – the Polish airline. Generally holidays were taken in company hostels in the mountains to the south and east of Poland or along the Baltic Sea coast.

The map extract of the Bieskady National Park (Figure 5.1) shows the very limited infrastructure for tourism in the Carpathian mountains in south east Poland, an area nominated for World Heritage status.

These mountains are the principal centre for skiing holidays in Poland. Although they are not very steep nor high, the low temperatures they experience between October and April create a six month season. Holidays in such a location were typical of what was available for Polish domestic tourism prior to 1990 (see Figure 5.2).

In the difficult transitional years after 1990, few people went away on holiday, as the number of tourists visiting Cracow shows (Figure 5.4). By 1996 there was a large increase. Poles had longed to be able to travel and in that year 23 per cent of the population, 7 million people, left Poland for a holiday abroad – mainly to Spain, Tunisia, Italy, Cyprus and Greece. A large percentage made their way to the cheaper Mediterranean resorts, where prices are relatively low and where the rather tired, decrepit accommodation was very acceptable after decades of hostel standards.

Cracow, a splendid mediaeval city in southern Poland (Figure 5.3), once the capital of the country, has a superb cultural heritage to offer visitors, albeit much still displays the bomb damage of the Second World War and is covered in grime from the nearby steelworks at Katowice.

The number of visitors increases annually (Figure 5.4), coming from countries such as Germany (13%), USA (11%), and Great Britain (8%). Tourism already employs 23 000 in Cracow, or 5 per cent of the total workforce. When the tourist infrastructure is improved, with better transport and hotels, the city and the mountainous region to the south will have attractions for thousands more visitors and will provide much needed employment and foreign exchange for the country.

FIGURE 5.2 Holiday accommodation available in the Bieskady National Park

FIGURE 5.3 Cracow

	Number of visitors to Cracow		
	Domestic	International	Total
1987	2 340 000	260 000	2 600 000
1991	1 376 000	42 4000	1 800 000
1996	2 700 000	1 000 000	3 700 000

FIGURE 5.4 Visitors to Cracow

At the other extreme there are many countries where state planning controls are almost non-existent or can be easily circumvented. Here private ownership and development is encouraged. Profit is the prime motive for the individuals, businesses and corporations who operate in the private sector. International companies such as the Sandals Resort Group, with resort hotels throughout the Caribbean, will use American finance, advertising and marketing to attract Europeans and Americans to their hotels. These capitalise on the warmth, scenery, beaches and relaxed life-style of the Caribbean Islands.

CASE STUDY

Sandals Resort, St. Kitts

Sandals resorts in the Caribbean are only open for couples. Bookings are not accepted for single people or family groups. Their main market is that of young couples from Europe or North America. However, as their brochure says: 'Once you experience Sandals you'll want to come back. Again and Again. Because once you enjoy a taste of Paradise, love will lead you back'.

Recently the company has developed a *Golden Players* club, which provides sports and fitness workouts for the 50+ age group.

They also offer a 'wedding moon' package with a marriage ceremony, personal co-ordinator, reception, bouquets, photos and video, honeymoon dinner and breakfast-in-bed the next morning.

Their clients come for a Caribbean holiday where accommodation is 5-star, sporting activities are included in the cost and there is no glimpse of hardship, poverty or discontent.

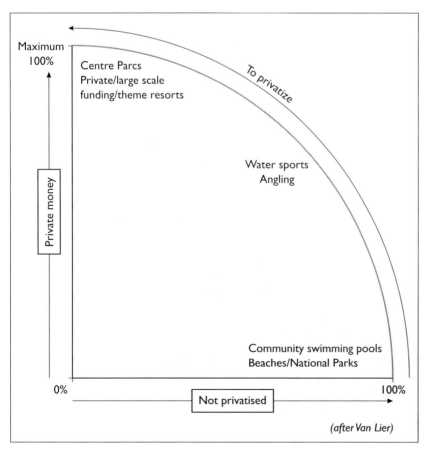

(after Van Lier)

Most countries practise a mixture of private and public ownership and development. Figure 5.5 shows how the range of public ownership varies from those recreational and tourist amenities provided and maintained by the authorities, to those whose development is entirely in the hands of private individuals and companies.

There has been a shift in attitude regarding the public responsibility for some areas of recreation. While care of public swimming baths, beaches, parks, forest areas and state buildings was once solely the responsibility of central or local government, nowadays this is devolved to private companies more and more. Most of the recent leisure parks built in Europe, North America, Japan and elsewhere are privately funded.

STUDENT ACTIVITY 5.1

1 With reference to the Poland and St. Kitts case studies, discuss what elements of tourist development should be the responsibility of the state.

FIGURE 5.5 The trend towards privatisation of leisure and tourism facilities in the Netherlands

The Management of Tourism

	Food	Housing	Expenditure (%) Health	Clothing	Travel	Other
Australia	23	24	6	6	12	29
France	21	18	14	6	12	29
Japan	23	19	10	6	14	28
UK	27	20	9	5	19	20
USA	14	20	14	6	12	34
West Germany	23	22	8	9	16	23

(Source: Wharton Econometric Forecasting)

FIGURE 5.6 Household expenditure: a comparison of major countries

Governments throughout the world are more immediately focused on the tangible benefits that accrue from tourism than on the needs of generations to come. As countries develop so their citizens have more disposable wealth and spend more of it on travel (see Figure 5.6).

The Role of Central Government

Travel expenditure represents important business for a government in the form of:

- Foreign exchange, if the tourist is from another country;
- Taxes paid by visitors, at airports etc.;
- Employment (5 per cent of the jobs in the EU states are tourist related);
- Economic diversification (tourism accounts for more than 4 per cent GDP in nine out of the 16 EU countries);
- Regional development;
- Capital investment.

It also brings a range of intangible costs and benefits. A visit to another country or a different region of one's own country, can enhance the products of that area long after the original trip. As people's travel horizons broaden, so their taste widens and they will seek those products in their home supermarket. Tourism also puts a place 'on the map'. Most governments recognise that sound control of tourism can bring good economic consequences. In particular they concentrate on the provision of the infrastructure of transport and public services, while leaving the construction of accommodation and facilities to private investment.

Another area of responsibility for the government is to pass laws relating to tourist control. Some countries still require all visitors to have an entry visa. Many limit the length of stay to six months. Travel controls during times of war or outbreak of disease are also generally imposed on the whole country. The Foreign Office in Britain will broadcast advice on when it is unsafe to travel to a particular country.

Executive Control

Various Ministries are responsible for different aspects of tourism; or in some countries where it is a dominant part of the economy, as in Spain, there will be a separate Ministry of Tourism. Decisions about the development of tourist resources will be made there.

Local Government

Some Regional or District Councils have wide powers to permit tourist development and to enforce various regulations. However, most of the countries of southern Europe have a strong central government and only limited autonomy, while in Switzerland a federal grouping of Cantons, or regions, means that most power stems from the regional capital. In Britain, county and district councils and urban boroughs have different policies towards tourism, depending on their individual needs. Each authority has a responsibility to grant planning permission for new buildings; in this way councils can control what sort of development takes place. Some wish to promote tourism. The outer London boroughs are keen to attract the international hotel chains who seek extra bed space near to London. Some wish to discourage tourism. Oxford, Cambridge and the London Boroughs of Kensington and Chelsea suffer from a surfeit of visitors.

Some towns have no need to promote tourism; they already have full employment. Other councils see tourist development as a last resort. Numbers employed in agriculture and other primary industries continue to decline; there is no manufacturing base and they are far from the 'sunrise strip', where hi-tech industries are locating. They hope that tourism will 'work for them' as it has in many other places. Tourism has a strong trickle-down effect: other industries will be attracted to an area that already has a positive image as a place to visit.

CASE STUDY

Ebbw Vale

Ebbw Vale is a town in South Wales that was closely associated with both the growth and decline of coal mining and steel making in the area from the mid eighteenth to the twentieth century. When the steel works finally closed in 1978, 32 years after the local coal mine had finished, the town suffered a period of massive unemployment. It is now beginning to show signs of economic growth again, as a rather unlikely centre for tourism.

In 1992 it was chosen as the site for the National Garden Festival as part of a deliberate policy of renewal. Large tracts of land were cleared of the debris of 150 years of coal, iron and steel production and were landscaped into an area of parkland. Millions of trees, shrubs and flowers were planted and the valley environment was transformed by introducing gardens, terraces, a waterfall and a lake. The site is now a permanent outdoors centre for visitors, known as Festival Park. It has a range of environments, from a Japanese garden to a tropical conservatory with children's activities and indoor exhibitions. Most recently a centre for factory outlet shopping has been opened on the site.

The town itself has turned the typical streets and shops of a Welsh mining village into a tourist attraction. It features as a stop on the *Valley Ways* motor trail, using the A465 Heads of the Valleys road as a link between different places and historic sites in the area. Two such locations are the Rhondda heritage park and Cyfarthfa Castle museum. Local breweries have been encouraged to provide some accommodation at suitable public houses along the route so that visitors can tour the area for a few days.

Ebbw Vale received an enormous financial boost when the Garden Festival was located there; subsequent developments show how tourism can successfully be used to regenerate a run-down industrial town.

The role of the European Union

Attempts to promote travel and mutual tolerance were encouraged throughout Europe after the Second World War in the wake of rapid growth of the mass market in tourism. The Helsinki Accord of 1975 refers to the need to facilitate travel, but tourism remained a low priority on the agenda until the 1980s, when Greece (1981), Spain (1986) and Portugal (1986) joined the Union. It is such an important facet of those countries that it developed a much higher profile almost immediately. The Maastricht Treaty of 1986 encouraged a dual approach to tourism through:

- **Horizontal Measures**. These aim to bring member states in line with each other so as to benefit from economies of scale. For instance, there are moves to stagger school holiday times throughout Europe so as to lengthen the holiday season. Another aim is to protect consumer rights by reducing bank charges for currency exchange.

- **Structural Measures**. These are initiated through the European Rural Development and Regeneration Fund. This channels grants into peripheral areas for specific projects. In particular the European Union has financed tourist projects in the border zone between Northern Ireland and Eire, in the Aquitaine area of France and in the Mezzogiorno in Italy.

Today tourism provides 5.5 per cent of the GNP of the European Union as a whole. It is responsible for 8 per cent of the goods produced and provides between 7.5 and 10 million jobs (from 4 per cent to 14 per cent in different countries). This discrepancy is due to the difference in definition of a tourism job in different countries. However, there is now a strong need for concerted action on behalf of those places where tourism is in decline before new zones of dereliction are created as happened in areas of mining and manufacturing decline.

STUDENT ACTIVITY 5.2

1   a) Summarise the reasons why most governments are keen to promote tourism.
b) Give evidence from any case study of how tourism can be used to revitalise an economy.
c) What other legislation do you think should be passed by the EU to promote tourism throughout Europe?

Models of Growth

Butler recognised five stages in his model of tourist growth and decline (see Figure 3.16).

In Stages 1 and 2 the area is discovered by travellers and explorers, whose pleasure lies in discovery and the opportunity of seeing other places and experiencing other cultures as they are. By Stage 3 all the amenities and facilities needed for mass tourism have been built. The resort is probably making a lot of money, but may be unrecognisable from the location first discovered 20 years previously. In Stages 4 and 5 tourism is in decline. This is particularly true of many of the Mediterranean resorts, where the hotels, shops and clubs that were built in the 1960s now look run-down, grubby and old-fashioned. These places find themselves in competition with newer locations such as Turkey, Thailand and Tunisia, where prices are lower and the surroundings are less commercialised and 'spoiled'.

It is up to governments at local, national and supranational level to reverse this decline. Many resorts are demolishing the high-rise hotels of the 1960s and are targeting the more affluent discerning traveller with a choice of smaller, luxury hotels and self-catering villas and apartments. Golf courses, marinas, horse-riding, swimming and fitness gyms are now all provided for the holiday maker of the twenty-first century, who no longer wants only to sit and sunbathe on the beach.

CASE STUDY

Magaluf, Mauorca

Magaluf is a holiday resort situated at Palma Bay on the south coast of the island of Mallorca, in the Mediterranean Sea, about 150 km off the coast of Spain. It grew as a resort area in the 1950s and has become associated with the cheaper end of the package business, accommodating thousands of visitors from the north of Europe, in high-rise hotels built along the sea front by multi-national companies.

In the 1990s many of these hotels have been demolished to create more open space. Replacement accommodation is further back from the shore-line and has been upgraded. The whole area is still very popular with tourists – 7.25 million people visited Mallorca in 1993. The tourism council of the island is making further efforts to improve the quality of the product on offer by the provision of a range of activity-related holidays, with golf, riding, sailing and cycling as popular pursuits. However, the increasing demand for water has led to problems of lowered water tables and increasing salinisation of farmland.

Magaluf now imports water by tanker from Spain. A recent proposal to build another golf course in the vicinity of the town has led to the growth of protest groups from among local residents. Their slogan *'s'aigna prima per als pagesos depres per als camp de golf'* – people come before golf courses – is an expression of the continued tension between the needs of the visitor and the resident.

Inequalities, North and South

There are various Development Models, such as that of Friedmann (see Figure 3.10) and that of Myrdal (see Figure 3.11) which set out alternative routes towards economic growth. The 'Tiger' economies of Asia based their growth on 'import substitution'. Local factories were established to manufacture goods that would otherwise be imported. Their products were then available for export to other countries and the country concerned started to earn the money needed for development into a modern state.

	Million baht
Manufactured Goods (largely jute/tyres)	266 148
Food	191 973
Machinery	176 465
Raw Materials (jute/rubber)	35 465
Tourism	110 000

FIGURE 5.7 Export revenue for Thailand

Other countries took another route to maturity. Thailand, for example, whose balance of payments depended on the export of primary products such as timber and metals, moved towards that of a mixed economy via tourism. Since 1983 it has been the fourth largest foreign currency earner in the country (see Figure 5.7). It has acted in a classic 'trickle-down' manner in that it has had a knock-on effect into agriculture, construction and import substitution.

Tourism is now seen as a way in which an LEDC can develop rapidly, particularly if it can attract investment capital from more wealthy countries. The power of tourism has also been recognised by world financial institutions aiming to assist in the growth of less developed countries. Many Caribbean states have received loans from American-based banks. For example, the Inter American Development bank loaned money to the Trinidad and Tobago government for tourist development, while the European Union has recently granted US$1 million to the island of

Grenada to establish a National Park. The European Union meets LEDCs at the Lomé Convention biannually, to discuss matters of common concern and interest. Recently tourism has reached the agenda and at the last meeting (Lomé IV) 9 million ECUs were allocated for tourism, specifically for 'human resources, market development and research'. The money may well be spent in Europe, where consultancy companies and universities will conduct market research, train and advise the national governments of LEDCs of the best way to develop their tourist industry.

Private investment tends to be more ruthlessly profit-focused than Government funding. Large hotel chains, airlines and tour operators will build hotels at suitable locations so that the advantage of the natural or cultural environment can be exploited fully. At times the power of capital is such that legislation, bylaws and court rulings will be circumvented to allow development to proceed.

CASE STUDY

Pattaya, Thailand

The resort of Pattaya in Thailand started life as a beach location for wealthy Thais from Bangkok. It was then developed with American capital during the period of the Vietnam war (1971–1975) as an 'R and R' centre (Rest and Recreation) for the US military. By 1978 independent Japanese consultants reported 'excessive pollution' of the water, the beaches and the land area surrounding the town. In 1983 the Mayor and the Council of Pattaya drew up a city plan in which 22 km of beach would be preserved from development. However, under pressure from the NESDB (National Economic and Social Development Board) and TAT (Tourism Authority Thailand), the plan was shelved.

By 1989 water shortage was so acute that some hotels received daily water deliveries by truck in order that their guests could wash. Water supplies to the townspeople were cut off completely.

Yet in 1990 Ambassador Hotels plc, a multinational company, received planning permission to build a resort hotel and convention centre with 3650 rooms and all the facilities associated with an international hotel chain. The town of Pattaya has grown by 50 000 since the early 1990s as migrants from the countryside arrive constantly, looking for work. They impose further strain on the already stretched resources.

The growth of such glittering monuments to western cultural norms imposes huge pressures on local customs, cultures and values. Local people seek employment in the hotels, generally at unskilled levels of work, while others produce cheap 'art and craft' souvenirs such as wood carvings and textiles. Some inevitably become involved in prostitution, crime, gambling and drugs. The beaches and forests will in some cases be reserved for tourists, while there will be rampant unchecked pollution of both land and water in 'native' areas. Many commentators see western tourism as the latest stage in a sequence of exploitation that began with colonisation; then imperialism followed by under-development and ultimate loss of control to MNCs.

Economic costs are also high. Imports of foreign goods are needed to meet tourists' tastes. Local people then buy these imported 'brand goods' in preference to their own indigenous products. Unless farmers are carefully directed, they may neglect to grow food for the visitors or for their traditional local markets, and then food supplies have to be imported.

Ultimately, a whole region or entire country may become dependent on its tourist industry as the main source of income. The market, however, is notoriously fickle and responsive to fashion and whim. Tourists can be easily dissuaded from visiting countries they believe to be unsafe. Terrorist attacks on tourist targets will destroy the industry for decades, as hotels in Northern Ireland, The Lebanon and Egypt have discovered. For many LEDCs the most recent threat comes from the current surge in 'cruising' as the fashionable way to go abroad. The visitor arrives by ship, spends a few hours ashore shopping or visiting the architectural highlights of the country (provided they are conveniently placed with respect to the port area) before retreating to the less-threatening world of cabins, sun-decks and restaurants aboard ship. The mid-centric tourist has found a way of taking home comforts with him or her, but that is little comfort for the hard-pressed LEDC!

STUDENT ACTIVITY 5.3

1 Use economic statistics for Thailand, The Gambia and other LEDCs to suggest how tourism can improve the standard of living in these countries.
2 Discuss the extent to which foreign investment in tourism in LEDCs is a form of neo-colonialism.

Managing Tourism and the Environment

Chapter 4 shows that tourists are particularly attracted to areas of natural beauty but that their presence can quickly lead to environmental degradation. The management of areas of natural environment is usually within the control of the authorities, whether they be local, regional or national. In Britain, a combination of government legislation and delegated authority is responsible for maintaining these areas.

Local Authorities were given power in the Countryside Act of 1968 'to provide, equip and manage' **country parks** within their areas for the provision of informal outdoor recreation for the general public. This has to be achieved within a framework of minimising visitor impact on the area concerned. The government also re-identified 33 **Areas of Outstanding Natural Beauty** or **AONB**s in the same Act, in order to preserve some of the loveliest parts of England from development. The Amenity Lands Act of 1965 protects beauty spots in Northern Ireland, while the Town and Country Planning (Scotland) Act of 1978 looks after National Scenic Areas in Scotland. There is no requirement to provide any recreational amenities at all in an AONB. Regions as diverse as the Gower Peninsula in Wales, the Logan Valley in Northern Ireland, Ben Nevis and Glen Coe in Scotland are included.

Protection of the environment can be extended further if it is felt necessary. The Agriculture Act of 1986 provided protection for 18 **Environmentally Sensitive Areas** throughout Great Britain, including Loch Lomond, The Broads and the South Downs. The traditional landscapes of these places are under threat from intensive farming methods. Farmers are given financial incentives to restrict their activities in the interests of preservation and conservation. Herbicides and pesticides are not used, hedges are retained and traditional wet-land areas are left to be wet! Each of these then provides an ecological niche in which a range of species of plants, insects, birds and animals are able to flourish. If absolute protection is needed, then a place can be declared a **Site of Special Scientific Interest** or **SSSI**, as stated in the 1981 Wildlife and Countryside Act. There are over 4000 such sites where there is a ban on any activity which would disturb flora, fauna, geological or physiographical features. These are areas where rare species are to be found and where any human activity is strictly controlled.

The case study of the Seven Sisters Country Park shows how a local authority can promote and enhance a classic landscape, which is also an AONB, to the benefit of both the public and the natural environment. Stages involve:

- an audit to discover the precise elements, both natural and man-made that are involved so that planning decisions can protect and enhance as required;
- a survey of local and national tourist and visitor patterns;
- an estimation of the 'carrying capacity' of the area;
- consultation with the community to define mutually agreed targets, maybe from conflicting opinions. These can include conservation, recreation, landscape preservation, the rural economy, support for farming, and reclaiming of derelict land.

CASE STUDY

The Seven Sisters Country Park

The Seven Sisters Country Park is situated in East Sussex, between Brighton and Eastbourne. It consists of an area of chalk downland behind the famous Seven Sisters cliffs, bounded by the River Cuckmere and Friston Forest (see Figures 5.8 and 5.9). The land was acquired by East Sussex County Council in 1971 and managed by Rangers. Its aims are:

- to provide opportunities for local people and tourists to enjoy the countryside;
- to conserve wildlife in a variety of habitats such as coastal, cliff, salt marsh, chalk downland and fluvial environments (see Figure 5.10);
- to preserve the outstanding scenic quality of chalk cliffs, chalk downland and the meanders on the River Cuckmere;
- to provide educational facilities and information for students and visitors to the area;
- to permit low intensity farming, generally cattle and sheep grazing, so as to preserve the traditional short grass landscape of chalk downland.

The impact of tourists and visitors to the country park is deliberately managed so as to cause minimum disturbance to the scenic quality of the area. Car parks are sited in nearby woodland or at the foot of the Cuckmere Valley, where the cars are screened by trees. New buildings are in local flint and old farm buildings are used to house tourist facilities such as the Visitors' Centre (Figure 5.11). Telephone kiosks are green instead of red.

FIGURE 5.8 The Seven Sisters, Sussex

- There are no litter bins in the park, so as to encourage people to take their litter home.

Visitors are also helped to enjoy the park as much as they are able:

- By the provision of a range of walks for different levels of fitness, some longer than others; passing through different habitats; offering a variety of views.
- By a programme of explanatory talks and walks for those who want to learn more.
- By linking the park to other tourist areas via signposts, tracks and roads. The South Downs Way, which is a long-distance footpath stretching from Winchester to Eastbourne, goes through the country park, and is clearly marked.
- By regular bus services to the nearby towns.

Rangers who were originally employed by the County Council are now part of the South Downs Conservation Service organise management of the park. It is a designated AONB (as formulated in the Countryside Act of 1949). It adjoins Lullington Heath (an SSSI because of the range of heath plants and insects to be found there). There have been suggestions that the South Downs should become a National Park. However, it is felt that the current management structure, in the hands of a joint body representing all the interested parties, is adequate protection for this outstanding heritage landscape.

Another management aim is to cause minimum disturbance to the wildlife and natural vegetation of the park.

- By provision of all-weather tracks away from the car parks so as to direct the majority of visitors towards zones of more robust vegetation.
- By signposting routes away from the most sensitive areas.
- By banning noisy activities such as moto-cross bikes.
- By the provision of picnic facilities, toilets and ice cream vendors close to each other so as to encourage people to eat in specified places.

FIGURE 5.9 Location of the Seven Sisters Country Park

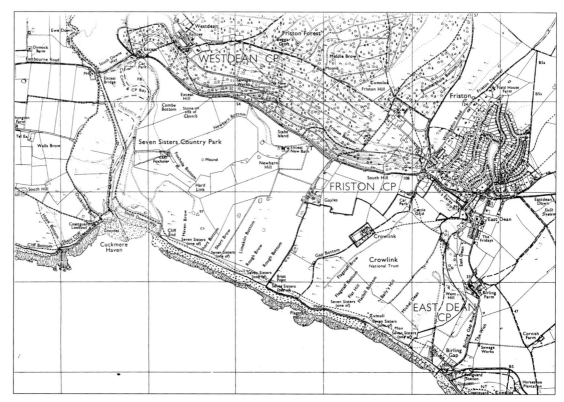

TOURISM AND MANAGEMENT 105

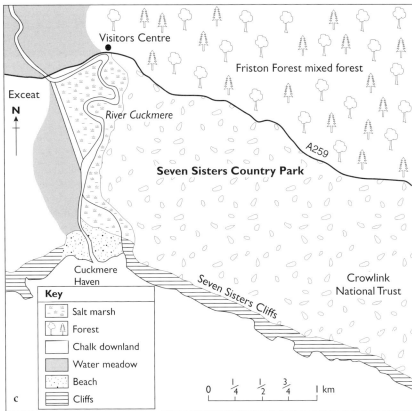

Figure 5.10a, b, c
a Saltmarsh along the Lower Cuckmere
b Visitors information board
c Habitats in the Seven Sisters Country Park

Figure 5.11 The Visitors' Centre, Seven Sisters Country Park

Conservation decisions at National level

Each country has to make its own decisions about the precise levels of conservation and commercialism within its territory. Most of the case studies in this book show elements of both protection and exploitation. Very often, in the case of a National Park, AONB or ESA in the UK, it is the protection afforded by an Act of Parliament that ensures the continuing success of tourist activity. Left to market forces, environmental degradation follows. Legislation also reflects the perspective and ethical stance of government and nation. In many countries wild game is controlled by hunting and this in turn is supervised by the issue of permits and licenses. Certain areas and times of year will be allocated to particular animals and birds. For example, it is not permitted to shoot grouse in Scotland until 12 August. This allows ample time for young birds to be hatched and weaned. There is a point of view that maintains that less damage is caused to a natural environment by controlled game sport than by banning it altogether in the name of species protection. Large numbers of jobs and income are also provided in areas which otherwise can be very poor.

National Parks

Most countries have passed legislation in order to protect the best of their natural environment. In 1864 President Lincoln declared the Yosemite

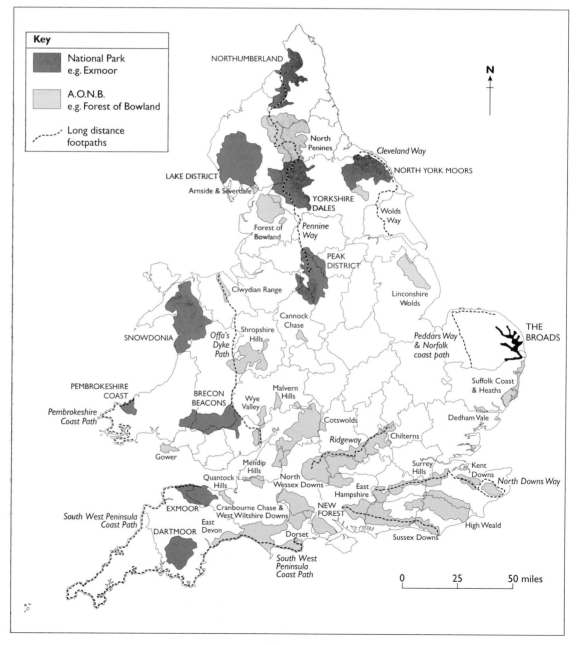

FIGURE 5.12 The National Parks, AONBs and Heritage Trails in England and Wales

TOURISM AND MANAGEMENT

Valley to be a public park 'inalienable for all time'. Then, in 1872, the Yellowstone National Park was the first area officially to be designated as national property. This was before the National Park Act of 1916, which put ownership and administration of National territories into government control.

In 1949 the National Parks and Access to the Countryside Act was passed in Britain to preserve and enhance the natural beauty of the areas specified. Ten parks were created (see Figure 5.12), covering 9 per cent of the land area of England and Wales.

The Act defined a National Park as being 'an extensive area of beautiful and relatively wild country in which, for the nation's benefit and by appropriate national decision and action, the characteristic landscape beauty is strictly preserved, access and facilities for public open air enjoyment are amply provided, wildlife, buildings and places of architectural and historical interest are suitably protected, while established farming is effectively maintained'.

Comparing the case study of the Smoky Mountains National Park in Chapter 4 with the following information on the Peak District shows that there are basic differences in the definition of a National Park from one country to another. In some, public access is strictly controlled in order to protect the environment, while in others, villages and towns make up part of the essential landscape and are included within park boundaries.

CASE STUDY

The Peak District National Park

The Peak District National Park is situated at the southern edge of the Pennine Hills and came into being in 1951 as the first to be created under the National Parks and Access to the Countryside Act of 1949. One third of the population of England and Wales live within 80 km of its boundaries so they are able to enjoy the unique environments created by the juxtaposition of Carboniferous limestone and Millstone Grit (see Figure 5.13).

Limestone underlies the White Peak area of the Park and creates a landscape of rolling hills and dales with the classic features of dry valleys, pot holes and bare 'pavement'. Millstone Grit is the dominant rock to the north, east and west, forming the area known as the Dark Peak. Its resistance to erosion means that it forms tracts of heather covered upland, edged by craggy, dark outcrops of grit. The quality of the natural environment is such that there are 21 National Nature Reserves and 51 Sites of Special Scientific Interest within its borders.

The Park covers nearly 1500 km^2 of territory, spread through six counties and is home to 37 400 people who live in settlements ranging from isolated farmsteads to small towns and villages such as Eyam, Dovedale and Castleton (see Figure 5.13). They follow a varied range of rural employment, for example tourism – 19.2 per cent, mining and quarrying – 14 per cent, water and energy supplies – 0.6 per cent. The average age of the population is increasing, as older people have bought property for retirement or because they can afford to commute out of the Park to places like Sheffield to work. Young people find it difficult to get jobs or housing in the area and have moved away to the cities for higher salaries and better facilities.

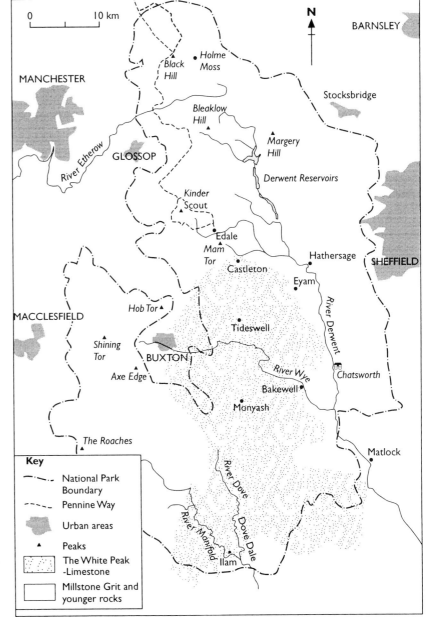

FIGURE 5.13 The Peak District National Park

In 1994, 22 million visitors were recorded as spending a day in the Park, coming to enjoy the landscape, to take part in the many outdoor activities, to visit one of the 30 historic Conservation areas or one of the 2500 listed buildings (the most famous of which is Chatsworth House). The sheer volume of their numbers causes immense problems to the Peak Park Management Board and to the residents. Beauty spots are swamped with people during the busy periods of the year, lanes and roads are choked with traffic to the extent that some have become one-way or traffic-free zones. Immense car parks disfigure the scenery, footpaths crumble under the onslaught of walkers, mountain bikers and horse riders. Residents dread the period between April and October!

Reconciling all these different interest groups to produce a successful rural policy for the area is not easy. Land ownership with its associated rights is illustrated below.

Water Companies	15%	The Peak Planning Board	4%
National Trust	10%	Forestry Commission	0.5%
Private Ownership	70%	Ministry of Defence	0.3%

It is important to maintain a balance of economic activities that are in many respects antagonistic to each other. Quarrying for road aggregate, limestone and gritstone is vital, both for the country as a whole as well as local employment. However, it ruins great tracts of landscape as well as coating surrounding areas with white dust. Demand for water from the catchment areas of the Howden, Ladybower and Derwent reservoirs restricts both farming and tourist development in their vicinity.

All these activities, along with the needs of both tourists and residents, need to be balanced if the unique landscape qualities of the area, which caused it to be selected as a National Park in the first place are to be preserved.

> **STUDENT ACTIVITY 5.4**
>
> 1 To what extent do you feel the provisions of the 1949 Access to the Countryside Act have been fulfilled by the creation of the Peak District National Park?
> 2 Reread the Smoky Mountain case study in Chapter 4 and discuss which form of management is most effective in preserving the environment as well as permitting public enjoyment.
> 3 Compare management provision of a local country park known to you with the Seven Sisters Country Park in East Sussex.

The development of ecotourism

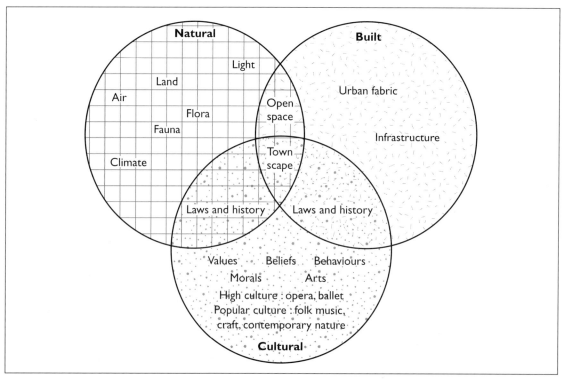

FIGURE 5.14 Elements of the Environment

(Source: OECD)

Environmental protection was a low priority throughout the years of mass tourist growth, both in northern Europe in the last decades of the nineteenth century and in the countries of the Mediterranean in the 1960s and 70s. In some areas tourist development became associated with landscape destruction, congestion and pollution. Both governments and private operators were interested in growth and profit regardless of the long-term damage to the natural environment and to the cultural and traditional life of the country.

The growth of organisations such as the *World Wildlife Fund*, *Greenpeace* and *Friends of the Earth* at this time, signalled the beginning of the revolt against the damage and destruction that was occurring. A group of concerned, powerful individuals from ten countries met to discuss the global predicament that non-stop expansion and mass production was creating. They issued a statement, now encapsulated in the phrase 'Limits to Growth' which suggested there was an alternative way forward. It is encouraging that there are signs of growing awareness, both from private companies and individual consumers, of the need to conserve and protect all the elements of the environment. The definition put forward by the OECD (see Figure 5.14) includes those facets of national life which give countries their individuality and which visitors hope to experience.

Government and organisations who provide grants for tourist development now do so in line with EIA (Environmental Impact Assessment) principles, the World Bank, European Union or other grant-making bodies are unlikely to finance projects whose effects will ultimately be destructive. The United Nations has taken a lead by its declaration of the need to conserve some of the most outstanding natural environments on earth as World Heritage Sites.

CASE STUDY

The Daintree River, North Queensland, Australia – a World Heritage Site

The Daintree River area of northern Queensland gained World Heritage status in 1988. It contains the last surviving area of untouched tropical rainforest in Australia, extending from the coral reefs, along the beaches to the granite peaks 1000 m above.

There is a unique range of plant communities including those of lowland montane rainforest, wet and dry sclerophyll forest, mangrove, fan palm and melaleuca swamps. Eight of the world's 14 primitive plant families have been found in the region. The associated fauna is similarly diverse. Some species, such as Bennett's Tree Kangaroo, are found nowhere else in the world. Bats, birds, butterflies and moths also contribute to this exceptional area.

Private landowners and the Queensland State Government were both anxious to sell off tracts of land to property developers. A road was constructed through Cape Tribulation National Park in 1983, despite protests from conservation societies and from the federal government in Canberra. Now that the area has received World Heritage status it should be easier for local politicians to stand firm against proposals for development. In 1993 an application that the electricity grid should be extended to the north of the Daintree river was turned down, despite immense pressure from Japanese resort development companies.

Sustainable Development

Tourist development that is sustainable in the future will have to focus on the needs of residents, visitors, tour operators and the environment. Local communities will expect to be consulted, maybe by using the Delphi Technique. In this process, local 'experts' from all walks of life are asked to:

- identify the most important considerations for the community in relation to the proposed scheme;
- rank the resulting list of issues that have been identified;
- comment on the ranked order that emerges.

In this way a consensus about the scheme, its strengths and weaknesses and whether it should proceed can be reached at an early stage of the planning process.

National and local authorities also need to carry out an Environmental Impact Assessment (EIA) survey. This has its origins in the USA, where it developed from cost-benefit analysis techniques first developed in the 1960s. Planners and developers would attempt to quantify the costs and benefits of a new scheme in financial terms in order to decide whether or not to go ahead. But there are always knock-on effects that cannot be quantified financially. How can you put a price on bird-song that is lost when woodland is destroyed? An EIA is a process that will identify the likely consequences both for the natural environment and for peoples' welfare of implementing new proposals. Figure 5.15 sets out the EIA headings for tourist development. Each of these is given a subjective score from −5 to +5. Both the total and individual negative scores will alert the planning authorities to unquantified costs.

FIGURE 5.15 A framework for developing an EIA

The natural environment.

1. Changes in the composition of plant/animal species by
   - Disruption of breeding habits
   - Killing of animals by hunting/to supply goods for souvenirs.
   - Affecting migration routes
   - Destruction of vegetation for building/for wood or plant gathering.
   - Creation of wildlife reserves.

2. Pollution caused by
   - Discharge of sewage/oil or chemical spillage into water
   - Vehicle emissions into air
   - Excessive noise from tourist activities

3. Erosion leading to
   - Soil Compaction which causes increased run-off/land-slides
   - Increased risk of avalanches
   - Geological damage
   - Damage to river banks

4. Natural resource depletion of
   - Ground water supplies
   - Fossil fuels to provide energy for tourism

5. Visual Impact affected by
   - The construction of facilities. e.g. hotels/chair-lifts/car parks
   - Litter

The built environment

1. Loss of land to
   - New construction
   - Altered drainage patterns
   - New infrastructure. Roads/railways/car parking grid systems/waste disposal sites

2. Altered visual appearance by
   - New building
   - New architectural styles
   - Visitors and their belongings

3. Changes in urban form due to
   - Changes in land-use e.g. residential houses become boarding houses
   - Changes to urban fabric e.g. more street furniture/signs/trees/flowers
   - Increased contrast between tourist and 'local' facilities
   - Restoration of buildings of tourist interest
   - Ultimate decline in interest in tourist facilities

Protection for the environment

With the best will in the world, it is still very difficult to protect an area from the impact of tourism. The Ecuadorian Government has done its best to protect the Galapagos Islands in the Pacific Ocean, as the following case study shows.

CASE STUDY

The Galapagos Islands, Ecuador

The government of Ecuador endeavours to maintain the fragile ecosystem on the Galapagos Isles in order to sustain the unique range of wildlife found there. These islands were first discovered in 1535 and subsequently used as a base by whalers and by buccaneers. Charles Darwin conducted extensive research into their fauna and flora as part of his famous expedition in the 'Beagle' in 1835. Shortly after that date they were annexed by Ecuador for use as a penal colony.

They represent a vital link both in the geographical spread of species between Asia and Latin America and in the chain of evolution first proposed by Darwin. Their unique importance was recognised by Ecuador's government when it declared the islands to be a National Park in 1959, and later extended its boundaries to include 50 000 km^2 of ocean for added protection. The Darwin Research Station at Santa Cruz is an international charity set up to help preserve species.

Tourism first developed in the 1960s and now there are nearly 60 000 visitors a year. People come from all over the world to appreciate the wildlife on the islands. The government protects with legislation, the trip is expensive, the tourists are committed to the sustaining of wildlife and yet the damage to the environment is increasing!

Today there are 30 hotels, two airports, two permanent cruise ships, many small boats and a resident population increasing at 10 per cent per annum, whose livelihood is closely related to tourism. The wildlife is disturbed by:

- tourists, whose mere presence frightens animals away from nesting grounds and usual habitats;
- goats, pigs and rats. These animals were introduced in the 1950s, either deliberately or by accident. It is now estimated that there are more than 50 000 goats and 1000 wild pigs living on the islands. They strip the vegetation and eat bird and reptile eggs;
- over-fishing. Seven million sea cucumbers were removed in 1994 although the quota is 500 000 per annum. (Sea cucumbers have aphrodisiac qualities!)

Every effort has been made to protect and preserve the unique fauna and flora of the Galapagos, both for its own sake and for our future understanding and research. Yet the advantages of awareness of past mistakes elsewhere, legislation by a sympathetic government and remoteness from the world tourist routes seem insufficient to protect these islands.

Protection for the visitor

Environmental policies can also protect tourists by setting standards of hygiene, health and safety. Probably one of the most effective measures in Europe was the European Union Directive of 1975, which set standards of safety for sea water and has gone some way towards control over the amounts of sewage dumped into the sea by local authorities and resort developers. The National Parks of Zambia also try to offer protection to the tourist by framing regulations to protect people in a hostile environment (see Figure 5.16).

Protection of the needs of the host community

It is important that the host community should accept whatever developments occur in their home area. Ideally, there should be some form of consultation and agreement beforehand, although one cannot always see the long-term consequences of decisions. For example, in the rural counties of Britain, planning permission to convert disused farm buildings into holiday cottages will generally be accepted as a valid alternative to disrepair and dereliction. When the visitors then discover that the landlord of the village pub provides tasty local cooking and spend each evening there, regular customers find that seats are taken and the atmosphere has changed. It is no longer 'their' territory.

It is also important that the local community should benefit economically from tourist development. The spread of safari parks onto native

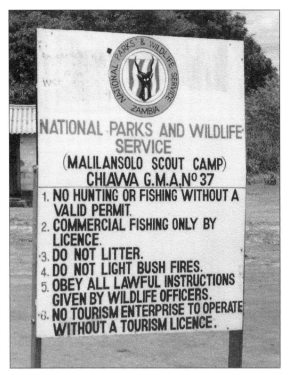

FIGURE 5.16 Regulations in Zambia's National Parks

farmland in Kenya has been well documented. Resentment from the villagers turned into criminal activity as the wildlife was then poached for skins, tusks and other saleable items. Elsewhere, various initiatives have been started by both governments and tour operators to 'share' the advantages of tourist growth.

CASE STUDY

Casamance, Senegal – West Africa

French tour operators have come to an arrangement with the villagers of Casamance in Senegal, whereby the villagers build and maintain one or two huts in their village for the use of holiday-makers, and undertake to act as hosts. They must provide food (native varieties) and activities for tourists who want a genuine experience of Africa. Only allo-centric travellers should apply! In this way the village receives the revenue that would normally go to the hotel company and the coach company.

CASE STUDY

Luangwa, Zambia

Elephants were rapidly overbreeding in the Luangwa National Park in Zambia, to the distress of local villagers. They showed their displeasure by illegal hunting and poaching. In this case, the authorities invited them to become gamekeepers. Villagers were employed to count herds, to track movements and migration patterns and to cull numbers when it was essential. In place of angry, resentful local people there were co-operative workers who could appreciate the value of the National Park now they were benefiting from it.

Future Directions

Although proponents of 'new' tourism write of a common purpose and focus whereby the industry works in harmony with the local community to protect the environment for the future benefit of all, there are in reality many unresolved conflicting interests still causing stress and tension in most tourist developments. The International Federation of Tour Operators (IFTO) has proposed the 'ECO Model of Sustainable Tourism' (ECOMOST), which sets out the following parameters for its members.

- The local population should remain prosperous and keep its cultural identity.
- The area should remain attractive to tourists.
- The economy of the area should not be damaged.

For those concerned with both community and environmental well-being, a more helpful list of objectives to promote sustainable tourism is suggested with the following pointers.

- Increase community involvement and planning.
- Establish carrying capacity levels.
- Combat pollution, congestion and degradation.
- Build in harmony with the physical and cultural environment.
- Protect the most vulnerable areas and peoples.

Those with responsibility for defining policy, planning and managing tourist developments should initially:
- define the nature and location of the proposed project;
- determine the overall purpose of the project;
- produce a strategic plan;
- raise awareness of economic impact e.g. income to GNP, foreign exchange, tax revenue, local employment and multiplier effect;
- focus on environmental impact e.g. levy on users to pay for conservation, measures to improve environmental quality, measures to reduce pollution and ecological disruption;
- protect the cultural identity of the host community.

Tourism will remain a growth area of the global economy for the forseeable future. As countries become economically more developed, so their citizens will have more disposable wealth and will choose to spend some of this money on travel. The travel industry will expand to meet this demand. It should, however, promote the interests of its customers and its raw materials alongside its profits and use the wide range of technology available to customise its product.

Butler's model of the tourist life cycle (page 64) focuses on the developmental stages of a resort. Plog (page 29) considers how the psychological make up of tourists dictates their travel patterns. These two models can possibly be combined in a holistic way. As people mature as travellers so they are prepared to go further from home and to experience different cultures and environments. This leads to the opening up of new markets as in Asia and Latin America. Meanwhile, those who are new 'global' tourists, from the countries of Eastern Europe and South East Asia are flocking to the popular resorts. The effect of increasing leisure time and more paid holidays from work is to increase the trend towards having more than one holiday a year. It seems that demand is infinitely elastic!

Special events, such as the Olympic Games and the World Cup, draw on a global audience for ticket sales and have turned the most unlikely destinations into tourist honey pots, while the impact of TV news coverage is such that individuals feel 'close' to tragic events and turn their locations into places of curiosity and pilgrimage. Supply is also infinitely elastic!

Maybe Pope John Paul XXI was right when he said that the ultimate result of tourism would be global peace and prosperity!

STUDENT ACTIVITY 5.5

1 The number of tourists globally is projected to grow exponentially until at least the year 2020 (Figure 2.4, page 29). Is it realistic to expect ecotourism to prevail?

2 Use the information in Figure 5.15 (Environmental Impact Assessment) to examine the effect of a recent tourist development in your local area.

3 Summarise the ways in which the governments of Ecuador and Zambia have endeavoured to protect both the environment and the tourist.

4 The case studies throughout this book illustrate a range of tourist development styles. To what extent do they reflect different perceptions of the national good by the states concerned?

TOURISM AND MANAGEMENT

EXAMINATION QUESTIONS

1 Illustrate the ways in which different national governments perceive and manage the social and environmental conflicts which can result from the development of tourism.
(50)
ULEAC

2 In what ways and for what reasons might people's attitudes and values towards tourism result in the development of different types of tourism and tourist facilities?
(50)
ULEAC

3 Briefly summarise how the economic gap between countries of the 'north' and 'south' has affected the pattern of tourism.
(10)
ULEAC

4 The growth of international tourism often causes problems and presents challenges in the tourist areas for planners and politicians. For areas you have studied, outline the nature of these problems and challenges, and evaluate attempts which have been made to respond to them.
(16)
Cambridge

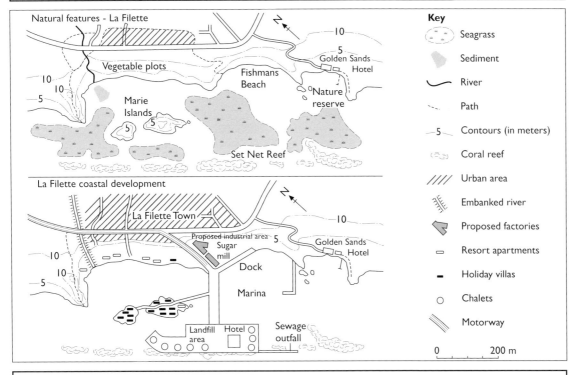

FIGURE 5.E.5 La Filette

5 a) Refer to Figure 5.E.5, briefly summarise, in a form suitable as a handout for the local press, the changes that may result from the La Filette coastal development proposals.
(7)

b) Evaluate the potential environmental and economic impacts of the proposals. You should consider the impacts on landform processes, ecosystems and existing and future economies.
(18)
ULEAC

The following examination questions could be answered with particular reference to one or more of the recurring case studies in this book.

1 Within the context of countries of the developing world, consider the extent to which the tourism industry is exploitative of the local environment and peoples, and has few positive effects.
(16)
Cambridge

2 For one region or country you have studied, examine the changes that have taken place in recreation and tourism in recent years.
(25)
AEB

3 In any one country or region you have studied, consider how the demands of tourism might be accommodated.
(16)
Cambridge

4 With reference to specific examples from case study areas discuss:
a the relative importance of the physical environment and local culture as attractions for international tourists.
(12)
b variations in and among the case study areas in the origin of international tourists.
(6)
c the factors which may discourage the growth of international tourism.
(7)
NEAB

Brighton

'In the next few decades tourism is predicted to become a major growth sector. It is therefore important that the South East exploits growth in demand for all types of tourism and the opportunity for substantial employment creation.

It is imperative that facilities and attractions are carefully managed taking into account sustainability objectives and respecting the local environmental and infrastructure carrying capacity.'

South East Structure Plan 1995

In 1995 the urban boroughs of Brighton and Hove were taken out of the East Sussex County Council administrative structure to form a separate unitary authority known as Brighton and Hove Borough Council. The reason for this move was to reduce the size and the population of the area under the control of the County Council into a more manageable unit and to recognise the separate urban identity of Brighton and Hove. The new authority has yet to produce its own development plans but others in existence, such as the County Structure Plan 1991–2011, the Coastal Strategy and the South East Regional Plan 1995 give clear indication as to the future direction of leisure and tourism in Brighton.

FIGURE B5.1 Tourist accommodation availability in Brighton

Type of accommodation	Number
5-star	200
4-star	557
3-star	835
2-star	
1-star	2000
Recognised boarding houses/guest houses	
Student Accommodation	2400

■ To enable all the residents of the town to have an opportunity to take part in a reasonable range of leisure activities, especially those characteristic of and compatible with the East Sussex environment.

– One of the features of Brighton mentioned by visitors in the 1973 tourist survey was the ease of access to the South Downs. These offer a range of leisure activities apart from the usual walking, pony trekking and mountain biking. Ditchling Beacon, which is situated immediately north of Brighton, has become a national location for paragliding and cycling hill climbs. Both these sports owe their existence to the tilted scarp structure of the chalk rock that provides steep slopes and thermal wind currents.

■ To develop and improve the range and quality of facilities and accommodation in order to maintain and wherever possible increase, the full economic potential of tourist and conference visitors.

– Figure B5.1 shows the number of bed spaces of different categories available in the town. There is little evidence to suggest this is inadequate, even when a major conference event, such as a political party's annual conference, is taking place.

– Brighton is well-supplied with bars, restaurants, clubs and cafes to meet the complete range of tastes. An eclectic variety of entertainment for all ages and outlooks is always to be found. It is this wide spectrum of facilities that earned the title 'London by the Sea'.

– Similarly, the range and variety of shops is typical of a much larger town and reflects the international cosmopolitan nature of many visitors to Brighton. This is a resource that needs careful strategic management if it is to stay vibrant and attractive. The development of more out of town shopping centres such as that at Holmbush (Shoreham), could easily reduce the centre of Brighton to a region of neglect and decay.

■ To encourage the provision of a range of facilities allowing diversity of experience and activity.

– Brighton came into being as a place where one could experience the delights of sea bathing! The sea is still available for the tough-footed enthusiast and in July and August the beaches will be crowded whenever there is a warm day and many people will be bathing in the sea, although none of the beaches has a 'blue flag'. There is also a designated nudist beach for those who wish to be uninhibited. However, there is no water activity attraction of the 'Wet 'n' Wild' variety. Such a venture would be immensely popular with many young people who go to Brighton, although the weather may be marginal for success.

■ To seek to make tourism and other leisure activities more sustainable by increasing accessibility by public transport, cycling or walking.

– Notwithstanding the large numbers of bus companies who compete to bring residents into the town centre for work, shopping or pleasure, and the main line rail station, which is only ½ mile from the central shopping area, Brighton has a major problem with cars. Parking space is inadequate for shoppers and discourages day-trip visitors to the sea-side. The park and ride schemes are aimed at residents.

– In its planning survey 'A Strategy for the Coast' produced by East Sussex County Council in 1995, reference is made to the successful bid for lottery money submitted for Millennium funding by Sustrans – the County

TOURISM AND MANAGEMENT

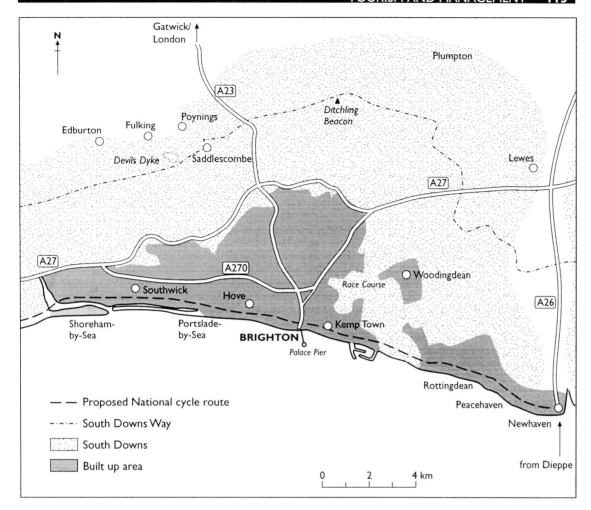

Council transport department. They propose a national cycle route right round the whole country that will pass along Brighton and Hove seafront (see Figure B5.2)
– Another tourist directed plan entitled 'Paths from Ports' proposes to link the South Downs Way, (a long distance footpath running along the crest of the Downs to the north of Brighton) with the Millennium cycle route, to and from the port of Newhaven. Visitors from abroad would be able to make a circular trip, on foot or by bicycle, through some outstanding scenery and would pass through Brighton en route (see Figure B5.2).

- To ensure development needs are balanced with proper concern for the East Sussex environment.
- To safeguard sites of national or historic interest.
    – Pressure groups, residents' associations, famous voices and national protection lobbies support both these aims whenever any new development is proposed in the vicinity of Brighton. A recent decision by Brighton and Hove Borough Council to allow the town football team to play in an existing stadium in the Borough, rather than expect the supporters to travel 130 km to a loaned ground, met with waves of protest from residents living nearby. They formed an Action Group and attempted to stop the move on legal grounds. Similarly, a proposal to turn a disused barn on the South Downs into an Environmental Studies centre was rejected on the grounds that the centre would inevitably disturb the environment it was intended to study!

The aims highlighted above come from the East Sussex Structure Plan 1991–2011 and are found in the section that focuses on Leisure and Tourism. The examples provided suggest it will be difficult both to safeguard the past and plan for the future in a way that meets with general approval in Brighton.

Spain

Although Spain was ruled by a strong centralist Government under General Franco from 1940 to 1978, little specific control or planning was directed towards tourism. Because of its rapid growth and the amount of foreign exchange it was generating, tourism was encouraged to flourish by the lack of regulations other than those which protected consumers and which categorised accommodation. Responsibility for tourism was passed from one government department to another, from Industry to Transport, until now when it rests with the Minister for Trade.

When the Spanish Constitution was rewritten in 1978 the country was divided into 17 regions, each with its own government, who separately were responsible for anything related to tourism other than overall policy (which was the concern of the Tourist General Board), and promotion abroad. TURESPANA co-ordinates the latter with an annual budget of 13 million pesetas. Their logo, designed by Joan Miro, hopefully encapsulates the image of Spain they wish foreigners to visualise (Figure S5.1).

Each of the 17 regions also promotes itself both internationally and internally by emphasising the specialities of their area. They are responsible for supervising and inspecting tourist business and activities. Provision of the infrastructure is left to individual town councils who are supposed to plan all road building, water, electricity and drainage provision within their boundaries. The separate forms of transport such as air (Iberia Airlines and several smaller companies), rail (RENFE and REVE) and sea transport (Trans Mediterranea) are all owned by the State.

Many of the problems faced by the Spanish tourist industry have come about through this division of responsibility. Planning is not a function of central government, but of local government who do not have sufficient power to make big decisions. During the late 1980s the Spanish tourist industry experienced a dramatic recession as numbers of visitors suddenly declined (see Figure S1.2).

This became the focus of the tabloid press in Britain with headlines such as 'Costa Sewage', 'Costa Stench' and 'Costa Nightmare'. Photos and stories about murdered waiters and drunken lager louts did nothing to encourage visitors to return. A Competivity Plan for Spanish tourism was drawn up by the Ministry for Industry to plan for the future and try to reverse the trend.

- New laws have now been passed to protect the coastal zone from further development.
    - The 'Ley de Costas' was passed in 1988, just before the economic downturn. It gives the public free access to beaches, dunes and cliffs not already lost to property development and prohibits building within 100 m of the shore.
    - Construction of new 1–3 star hotels has been banned throughout the established tourist areas of the country.
    - VAT has been reduced from 15 per cent to 6 per cent on 5-star hotels in order to attract more high-spending tourists.
    - Financial regulations have been changed.
    - Devaluation of the peseta on four occasions between September 1992 and March 1994 allowed Spain to retain the package holiday trade at the cheaper end of the market, particularly in view of the potential number of visitors from Eastern Europe. (The Moscow Consulate issued 80 000 visitors' visas in 1994.)
- Modernise facilities throughout Spain.
    - New buildings are more strictly controlled than in the past. Many of the shabby, high-rise hotel blocks have been demolished.
    - Road building now concentrates on by-passing some of the most popular resort sea-fronts.
    - Water, sewage and electricity supplies have been upgraded.
- Restore the Environment.
    - £65 million has been spent in the Balearic Isles on waste disposal, road lighting, beach replenishment and park construction.
    - Control over sewage emissions into the sea has improved. The incidence of coliform bacteria in sea water has been reduced. Spain now has 229 Blue Flag beaches (by comparison with 20 in Britain). In a survey of German tourists conducted by Neckermann, the German tour operator, 60 per cent of those who responded said that clean water and clean

FIGURE S5.1 Spanish Tourism logo

FIGURE S5.2 Sludge spill

Spanish mine waste warning 'was ignored'

Spanish authorities were yesterday being blamed for the wave of toxic sludge from a mine reservoir that is threatening marine and bird life in Europe's biggest natural park, the Coto de Doñana in south-west Spain.

Environmentalists said authorities had been warned several times of the dangerous state of the reservoir dam, but all had considered the mine and its reservoir, owned by the Canadian Boliden group, to be safe. In 1996 both the government of the southern region of Andalusia and a local court at Sanlúcar de Barrameda rejected complaints from a former mine manager that the reservoir was a danger to Europe's most important area of wetlands. Manuel Aguilar Campos, a former mine manager, had said that the retaining walls of the reservoir near the town of Aznalcóllar, were ready to burst.

Boliden blamed the spillage of the five million cubic metres of highly acidic, sulphurous waste on an underground earth movement. It said it would meet any legal obligations resulting from the spillage.

Dozens of dumper trucks were at the mine yesterday, transporting rubble to fill the 150ft breach that opened in the dam on Saturday. The mine and its reservoir, which tower above the River Guadiamar on the outskirts of Aznalcóllar, will remain closed for at least six months.

Yesterday the toxic waste appeared to have been diverted away from the most delicate wetland zones of the Doñana National Park. Officials said it should make its way into the sea over the next few days. But a former senior park official said that rainfall could still flood the park with sludge deposited farther up the foul-smelling river.

Greenpeace said the huge tide of poisonous waste was killing everything in its path as it moved downstream in rivers and man-made channels to the Gulf of Cádiz. Isabel Tocino, the Environment Minister, said that the Doñana reserve was out of danger, but she described the ecological damage to the region as 'catastrophic'.

Scientists declared that the heavy metals contained in the waste water, including zinc and lead, had already started to make their way into the food chain.

Dead carp lay on the thick black sludge that extends for up to half a mile on either bank of the River Guadiamar. Groups of storks could yesterday be seen wading into the toxic mud to eat the poisoned fish. In some areas park officers used gunshots to scare the birds away.

Scientists said that a multitude of crabs and small crustaceans would also have been killed by the toxic waste and that these would be eaten by some of the quarter of a million birds now nesting in Doñana.

'Our worry is that birds will now come to eat the dead creatures and that they, in turn, will start to die in large numbers,' said Juan Carlos del Olmo of the World Wide Fund for Nature.

Shrimp and eel fishermen in the tourist town of Sanlúcar de Barrameda, at the mouth of the River Guadalquivir, said the diverted toxic waste would inevitably leave them without a livelihood.

'They care more about ducks than they do about people,' one complained.

A Spanish farmers' group yesterday put damage to crops resulting from the spill at a preliminary 1.5 billion pesetas (£6 million).

The Times, 28 April 1998

beaches were their first priority when choosing a sea-side holiday location.
- Nature reserves. There seems to be less protection of the natural environment away from the coast, particularly near nature reserves and National Parks. In the case of the Cota Donana reserve on the Guadalquivir river in south west Spain, extraction of water for nearby agricultural land damaged the wetlands, on which the reserve is based, and disturbed the delicate ecological balance of the region. A recent spillage of waste water from a mine reservoir near the town of Aznalcollar in Andalusia, threatened the region once again (see Figure S5.2).
- Offer new products to tourists. The Ministry has made a strong effort to broaden the base of tourism by:

- encouraging the construction of retirement property around Alicante and the Costa Del Sol so as to reduce the effects of seasonality;
- promoting rural, inland tourism by constructing modern roads to inland towns such as Mijas (a hill-top village) and Ronda (famous for its ravine and bridge), which are now the focus of day trips from the coastal resorts;
- widening the range of inland recreational activities and sports facilities (see Figure S5.3);
- promoting thermal spa resorts and talassotherapy (water therapy) at both inland and coastal resorts;
- promoting cultural and urban tourism (this is detailed in the Spanish section in Chapter 4).

All of these efforts appear to have been successful. The attempt to improve the quality of accommodation on offer led to an increase in tourists from the United States:

| 1994 | 997 000 visitors from USA | 27.2% increase on previous year |
| 1995 | 1 600 000 visitors from USA | 62.8% increase on previous year |

Other arrivals show a similar increase in numbers for 1995:

20 300 000	visitors from France	48.1% increase on previous year
10 400 000	visitors from Germany	7.9% increase on previous year
8 800 000	from Britain	4% increase on previous year
3 700 000	from Italy	30.8% increase on previous year

Butler's model of tourist growth offers a variety of outcomes once stagnation has been reached. Many other countries will be watching to see if Spain has managed to reverse the decline in numbers of visitors experienced in the late 1980s. The alternative to success is the creation of regions of economic depression similar to those experienced throughout northern Europe in the steel and coal mining industries. All of Europe hopes this will not be so.

FIGURE S5.3 Activities in rural Spain

ACTIVIDADES

El Valle de Tena ofrece uno de los más completos y variados programas de actividades en alta montaña. Actividades de todo tipo, al alcance de cualquier nivel, desde debutantes a iniciados, para los más tranquilos o para los más audaces, dan la oportunidad de disfrutar de todos y cada uno de los rincones del valle, al estar distribuidas por toda su geografía.

Deportes de aventura

En los que los amantes de la emoción disfrutan al límite de sus posibilidades. Descenso de barrancos, una nueva modalidad dentro de los deportes de montaña que cada día cobra adeptos. Escalada en paredes artificiales (rocódromo) y en las paredes naturales del valle, completamente equipadas. Rafting, siguiendo el curos del río Gállego desde Oliván hasta Sabiñánigo.Además de espeleología, parapente y travesías de montaña.

Actividades náuticas

En el pantano de Búbal, rodeados por un magnifico paisaje, cursillos de wind-surf, piragüismo, vela ligera y alquiler de embarcaciones e hidropedales, en una zona de recreo con servicio de bar y cómodas hamacas para tomar el sol.

Excursiones

A pie, a caballo, en bicicleta de montaña en vehículos 'bodo terreno', en telesilla o en helicóptero, dan a conocer, desde diferentes prismas, la inmensidad del valle (Se facilitan planos de recorridos). Como novedad un tren turístico descubierto le llevará a conocer fantásticos pasiajes.

Actividades didácticas

Existen diferentes empresas formadas por profesionales cualificados, que le pueden acompañar e iniciar con seguridad en todas las actividades.

 ARAGON AVENTURA
Somos especialistas en actividades de montaña:
Escalada, Alta Montaña, Descenso de Cañones,
Rafting, Multiactividad en Montaña.

The Gambia
Planning for the Future

It is difficult to see any way in which The Gambia can develop rapidly from its very low base, as fifteenth poorest nation in the world, with a per capita income of only US$230 per annum. The country is heavily dependent on agriculture. Eighty per cent of the population is involved in some aspect of farming. The men mainly plant the cash crop of groundnuts while the women grow the food to feed the family. This consists of planting vegetables and cereals, millet and sorghum on the drier land away from the river, and rice on land that can be watered or irrigated.

Recently, a scheme to increase family income by encouraging women to grow vegetables for sale, both to the hotels along the coast at Bakau and for export through Yundum Airport, has been very successful (see Figure G5.1).

Loans from the UNDP (United Nations Development Programmes), the EU and the IDB (Islamic Development Bank) of £1000 per village for land clearance and provision of fencing, tools, seeds and wells to each group of women, enabled them to plant additional crops on newly established plots. These were then sold to a Wholesale Company named Citroproducts (a government subsidiary) who marketed them on behalf of the village co-operatives. The villagers quickly repaid their loans and during the 1990s established a regular trade to Europe of aubergines, chillies, okra and green beans.

The traditional role of men in west Africa is that of herder and farmer of cash crops, in the case of The Gambia this is groundnuts. Currently the market is very unstable, being subject to the basic laws of supply and demand, so that prices for groundnuts fall in a good year. In addition they have to compete with a wide variety of vegetable oils, such as oil seed rape and sunflower oil, which can be grown in Europe.

There are many NGOs (Non-Governmental Organisations) such as Oxfam, Action Aid and CAFOD working with the Gambian population hoping to find ways to diversify income, particularly in rural areas, so as to improve the standard of living for everyone. Their schemes are usually based on an extension of rural activities and look towards provision of roads, causeways, wells and machinery on a credit basis. Much of what is produced could be sold to the hotels, particularly if farmers and fishermen are encouraged to produce European varieties.

Inevitably, young people leave the countryside with its limited range of opportunities and make their way to the towns where they look for employment.

It is difficult to see how the proportion of income from tourism can be radically enlarged. Outside of the coastal areas the roads are only surfaced with laterite and there is only a very limited range of potential activities and sights to be seen upcountry. It would be possible to build a few more hotels along the coastal strip, although even these remaining undeveloped hectares are very vulnerable to erosion by the strong currents around the estuary.

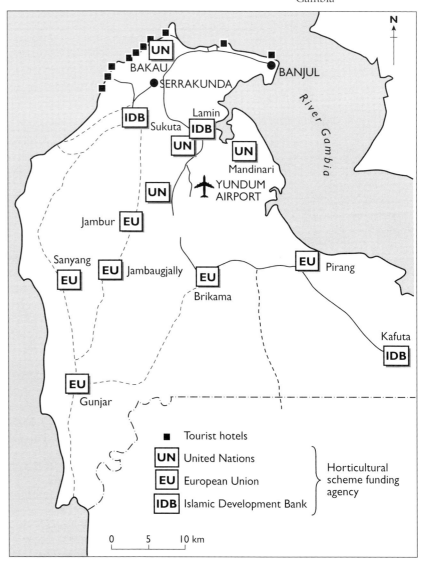

FIGURE G5.1 Village agricultural co-operatives in The Gambia

Maybe The Gambia can find a niche in the provision of specialist holidays for a discerning minority of people. Its location on the main bird migration routes between Europe and Africa and the wide range of tropical habitats found within the country – mangrove swamp, rain forest and savannah grassland – mean that it is possible to see a large percentage of the birds of both continents in one place. Bird watching holidays already exist and could be developed further.

Overall it may be best for economic development to take place within some form of federation with other West African states, especially Senegal. In Senegal the tour operators have linked up with individual villages in Casamance Province to provide tourists with a genuine experience of African life. Senegal could also provide The Gambia with a market for its produce. Certainly the future of The Gambia lies in development of the villages and countryside in as many ways as possible. Tourism, either in search of sun, sea and sand or for birdwatchers, can only provide a small proportion of the income and employment needed by this tiny state if it is to raise its standard of living.

New Zealand

FIGURE NZ5.1 The number of international visitors to New Zealand

1991	963 000
1992	1 056 000
1993	1 157 000
1994	1 323 000
1995	1 409 000
1997	1 541 136

As with most other developed countries, tourism plays an important role in the economy of New Zealand, contributing nearly 14.5 per cent of export earnings and providing employment for a significant proportion of the national workforce. In some areas such as Westland, South Island and the Rotorua area of North Island, it is the dominant industry. The central government in Wellington has always recognised its value and as early as 1 February 1901, the first Tourism Department was established within the Railways Department as 'part of the scheme for utilising the railways to a greater extent than is now done in popularising the sanatoriums'. In this respect the growth of tourism was seen as part of the executive responsibility for health (by promoting the thermal resorts) and transport.

The twentieth century has seen continued government involvement in the promotion of tourism to a greater or lesser extent. In many ways the division of power reflects that established in Great Britain, from where most of the early legislators came. Central government still plays a strong role in overall planning and strategic development, while local authorities are responsible for land-use decisions. Deregulation and privatisation in the 1980s led to the creation of independent transport companies such as New Zealand Airlines and to the growth of foreign investment in hotels at the major resorts. However, much of the provision for tourists still rests with family-owned and run businesses, guest houses, motels, cafes and boat yards etc.

The New Zealand tourist department redefined its objectives in 1990 'to develop and market New Zealand as a tourist destination where this is beyond the interest of the private sector and where this is a cost-effective solution to the government's desired outcome'. It aims to promote New Zealand abroad as a destination offering:

- world famous scenery;
- uncrowded, wide open spaces;
- the personality and culture of friendly, welcoming people;
- the opportunity to participate in outdoor activities;
- a remote, exotic location.

Its target audience has been identified as those sectors of the market that are most likely to travel to New Zealand – the older and younger spirited travellers. The plan is to reach these groups of people as frequently as possible, but not to bother with advertisements aimed at other groups.

Internally, the department endeavours to broaden the spread of the industry by promoting towns and regions outside of the 'Big Four' (Auckland, Rotorua, Christchurch and Queenstown) and the 'blue riband route' (see figure NZ5.2). This is particularly important for the peripheral regions such as Westland in South Island and the far north of North Island.

The peninsula to the north of Auckland is known as Northland and because of its relatively mild climate, is often referred to as the 'winterless' north. It is a popular holiday resort area for New Zealanders, but has relatively few international visitors (20 471 in 1995 compared to 696 000 domestic holidaymakers). The main emphasis is on water sports off the coast at areas such as Ninety-mile Beach and the Bay of Islands (see Figure NZ5.2). However, there are relatively few alternative economic activities apart from farming and fishing and the resident population numbers only 136 800 (54 400 in the North District, 64 900 in Whangerei District and 17 450 in Kaipara District).

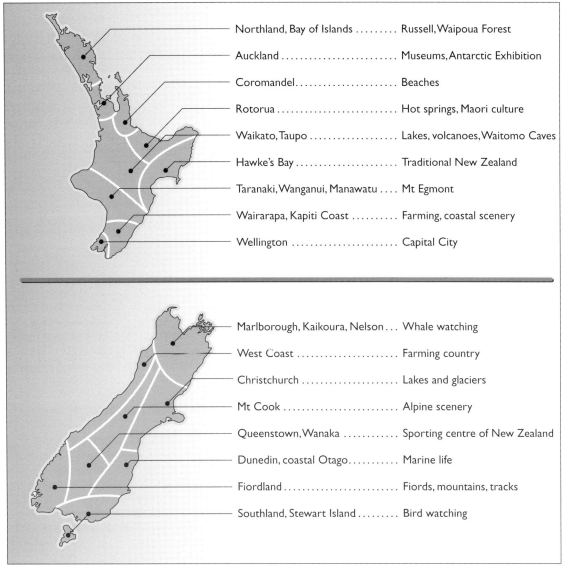

Figure NZ5.2 The main tourist areas in New Zealand

(Source: New Zealand Tourist Board)

It was at Waitangi, Kaipara, that the first European settlers talked to the Maori residents and signed the Treaty in 1840 which ceded so much land into European ownership. Maoris make up 26 per cent of the population of Northland, a larger proportion than elsewhere in the country. They feel strongly that they should have a corresponding share in the economic development of the region.

A recent survey of small businesses in Northland which provided services to tourists, such as cafes and bed and breakfasts revealed that 11 per cent were Maori, owned (approximately 28 businesses). Of these, two were owned by extended family groups, one by a Maori Trust fund and one by a tribal group. The rest were individually owned. These economic ventures, like most small businesses world-wide consisted in the main of one or two individuals investing their savings in an activity at which they hope to make a living and a profit.

These owners have few advanced business skills or assets other than their capacity for hard work. They are vulnerable to trends and market forces outside their control, such as shifts in the exchange rate and periods of good or bad weather, to determine their success or failure.

Maoris make up 19 per cent of the employees in these small businesses in Northland and it is this underlying fact that causes their leaders so much concern. They wish to share in the economic growth of the region as equals, but seem to be cast in the role of dependant.

Hall, in his book *Tourism and Indigenous People* writes 'Maoris are successfully reclaiming some of the most famous land areas and landmarks of New Zealand on the grounds of their prior ownership and of the dubious legality of some of the early land sales to Europeans. If they are to cope successfully with the future development of tourism in these areas then they will need business skills of a high order. Their cultural identity may be diluted or even lost in the process'.

The Disney Concept

FIGURE D5.1 Attendance at North American theme parks

Rank	Amusement/Theme Park	Attendance (millions)	% Change 1995/96
1.	Disneyland, Anaheim, California	15.0	+6%
2.	Disney's The Magic Kingdom, Orlando, Florida	13.8	+7
3.	Epcot at Disney World, Orlando, Florida	11.3	+5
4.	Disney–MGM Studios, Orlando, Florida	10.0	+5
5.	Universal Studios Florida, Orlando, Florida	8.4	+5
6.	Universal Studios Hollywood, Los Angeles, California	5.4	+15
7.	Sea World of Florida, Orlando, Florida	5.1	+3
8.	Busch Gardens, Tampa, Florida	4.2	+10
9.	Six Flags Great Adventure, Jackson, New Jersey	4.0	Flat
10.	Sea World of California, San Diego, California	3.9	+4

Attendance figures are based on estimates from magazine sources as reported in Travel Weekly.
Source: Amusement Business Magazine, Nashville (TN)

The 1996 attendance figures for North American theme parks (Figure D5.1) suggest that the Disney Corporation is alive and well!

Encouraged by the continual popularity of their enterprises in Florida, the Corporation opened the Wild Animal Kingdom in 1998 to offer an alternative experience to their visitors.

Other companies are competing to maintain their share of the market, for example, a planned US$2.5 billion expansion to the Universal Studios theme park in Orlando, which will create a 'stand-alone destination resort'. Visitors to the theme park will have no need or desire to leave the premises for the entire period of their stay! There will be 4700 hotel rooms, a night-time entertainment district and a golf course.

Sea World of Florida has added new attractions to its existing park and is building a new park to be known as 'Key West World' next door.

In Europe the Disney Corporation is finding it much harder to match the performance figures of its American parent company. In 1998 there was a decrease in the number of visitors to Eurodisney, attributed to the poor early summer weather and to the counter attractions of the World Cup in France. However, the complex is proving to be an attractive site for business conventions and as a conference venue. The number of hotel lettings for that purpose increased as 40 per cent more delegates booked in to use a second conference centre, recently opened on the site. Although 1998 sales figures are 8.4 per cent up on the previous year's, this came largely from the sale of shop leases in the recently opened 'Val d'Europe' shopping centre. Every available scrap of income will be required to meet the £30 million loan interest charges due on a debt of £1.5 billion.

The Corporation's plans for the future remain secret, as they would not wish to inform the opposition of their intentions. However, one development is already sufficiently advanced to be general knowledge. Disneyland is entering the cruise market with two ships initially, together carrying 1740 passengers. These will operate out of Fort Lauderdale, Florida, and will be part of an inclusive package sold to visitors. A one-week holiday will include 3–4 days in the theme park and the remainder cruising at sea. Facilities on board will be provided for a younger clientele than are normally catered for in the cruise market.

It is evident from these case studies that the Disneyland Corporation is very aware of the Butler model of the tourist resort life cycle and is determined to reinvent and rejuvenate its product rather than face stagnation and decline.

INDEX

Abuko Nature reserve 87–8
Accommodation 32
Activity holidays 13–15
Acts of Parliament 101
Airlines 56
Alfriston 41
Alpine tourism 76
Althorp 80
Amenities 40–47

Balance of payments 60
Barcelona 85
Beach cleanliness 77
Biomes 74
Brighton
 – amenities 45–6, 65
 – conservation 84
 – history 19
 – development plans 112–14
 – morphology 44–6
Built Environment 71, 113
Butler model 64, 116

Communications 32
Communism 93–6
Community issues 26–8, 49–50, 52
Conference centres 65
Conservation/nature 80, 104
Conservation/urban 83
Consumer profiles 69
Cost/benefit 107
Country Park 101
Courchevel – France 38
Cruises 12, 120

Daintree river site 107
Degradation 75, 107, 115
Deregulation 54–6
Destinations 6, 13, 17, 29
Development plans
 – Brighton 112
 – New Zealand 118
 – Spain 114
Diversification
 – Spain 116
 – Gambia 117
Disney concept
 – animal kingdom 91

 – community impact 52
 – development plans 120
 – early growth 25
 – Eurodisney 70
community impact 52
Ebbw vale 98
ECO Model 110
Economics 53–8
Ecosystems 72–6
Egypt 16
Employment 7, 63–7
Enviromental
Impact Assessment 108
Env. Protection 101
Euro Disney 70
Euro-tunnel 12
European Union 98
Events 17–18, 42–3, 86
Exchange rates 17
Expenditure 53
Exploitation 26–7

Flows 14, 17, 29
Friedmann model 61

Galapagos Is. 108
Gambia (see The Gambia)
Gatwick Airport 31
Gender 26–7
GNP 59
Government role 97
Gracelands 80
Gt. Smoky Mts. 81

Historical dev:
 – Brighton 19
Hot-spots/honeypots 15, 41, 77

Inequalities: North & South 99
Income growth 7–8, 10
Indonesia 56
Infrastructure 62

Landscape
 – natural 71–4
 – New Zealand 23–4
Leakage 63
LEDCs 99

Leisure
- definition 4
- time 40
Liverpool 34

Mallorca 39
Management of
 ecosystems 73–5
Management of
 tourism 97–101
Maoris 51, 119
MEDC 60
MNCs 59, 100
Models 61, 64
Morphology of
 resorts 44–7
Multiplier 61, 63, 68
Myrdal 61

National Parks 81, 104, 106
Zambia 109–10
Poland 94
National Trust 72
New Zealand
- consumer profiles 69
- ecotourism 89
- Maori role 51, 118
- natural landscapes 23–4
- Nature reserves 87, 115

Olympics 43, 86
Ownership of
 resources 15, 96

Package holidays 9
Patterns of travel 29
Peak District 104–5
perceptions 93
Philosophies 16, 93
Plog. Psychology of
 tourists 29
piers 84
Poland 94–5
political influences 16
protection 109

Recreation
- definition 4
- resources 15, 72
Regeneration
- urban 34, 85
- rural 36
Regional
development 61
Resorts 38, 42–48

SAGA holidays 57
Savannah grasslands 75

Second homes 35
Senegal 110
Seven Sisters Country Park 104
Sheffield 40
Social issues 26–7, 43, 49, 51
Social patterns 7, 11
Spa towns 35
Spain
- development 114
- employment 66
- migration 66
- regeneration 85
- reasons for growth 20–21

St. Kitts (Sandals) 96
St. Tropez 15
Stonehenge 79
Sustainable tourism 71, 107, 110
Sydney, Olympics 43

Thailand 99–100
The Gambia
- diversification 117
- employment 67
- Nature Reserves 87
- reasons for growth 22
- social issues 49–50
Theme Parks 25, 42
Tour operators 9, 14, 55, 67
Tourism Definitions 4–5
Historical development 9–10, 58
Tourists, expenditure 7
- types 17
Transit operators 56
- Transport 7, 30, 62
- air 13, 30–1, 50
- car 12
- coach 12–13
- ferry 12
- rail 11
Travel agents 32, 55
Travel times, costs 8, 30

Urban regeneration 34

Victoria Falls 78
VFR tourists 5, 68

Weather-impact 15, 16
Whale spotting 81
Work patterns 10
World heritage site 107

Ynyslas sand dunes 73
York 28

Zambia 109–110
Zimbabwe 61, 75